THE JETS AND SUPERCRITICAL ACCRETION DISK IN SS433

Contents

Introduction	3
Parameters of SS433	5
The jets of SS433	6
The optical jets	9
The jet spectral lines	9
Kinematic model and precession of the jets	12
How the moving lines move	16
Geometric and kinematic parameters of the jets	19
The radio jets and W50	23
The uniqueness of the SS433 radio jets	23
Radio variability	26
Flares	29
Radio brightening zone	32
Equatorial wind	35
W50	37
The extended jets	39
The X-ray jets	43
Early observations	43
Localisation of the X-ray source	44
ASCA data. Lines and spectrum of the jets	46
ASCA data. Equatorial wind	48
CHANDRA data. Narrow multi-temperature jets	49
Inhomogeneity of the jets and X-ray variability	55
Structure and formation of the jets	57
The state of the gas in the optical jets	57
The zones of sweeping out and expansion	62
Heating of the jets	65
Ejection of gas in the jets	69
Acceleration, collimation and fragmentation	73

The supercritical accretion disk and the components from photometric data	80
The light curve of SS433: precessional, orbital and nutational variability	80
The light curve in active and quiescent states	86
The nutational clock and time for the passage of material through the disk	87
Outflows in the disk plane and gaseous flows	92
Sharp brightness decreases	94
The spectral energy distribution and parameters of the components	95
Polarisation of the optical and UV radiation	98
The supercritical disk from spectroscopic data	100
The "stationary" spectrum of SS433	100
He II radial-velocity curves and the mass function	101
The component mass ratio	104
Properties of the gas flow from the He II and Hβ lines	106
The structure of the disk and the central region from the He II line	108
Precessional modulation of the stationary lines	111
Variability of absorption lines. Profile of the velocity of the disk wind	112
Structure of equatorial outflows in SS433	116
Gas streams in SS433	120
SS433 and microquasars	122
Microquasars	122
Supercritical transients	126
A Face-on SS433 and ultraluminous X-ray sources in other galaxies	129
Acknowledgements	137
References	137
Index	151

ASTROPHYSICS AND SPACE PHYSICS REVIEWS

Editor: R.A. Sunyaev
Institute for Space Research
Russian Academy of Sciences
Moscow, Russia

Advisory Editor: M. Longair
Cavendish Laboratory
University of Cambridge
UK

GENERAL INFORMATION

Aims and Scope
Astrophysics and Space Physics Reviews publishes review articles covering significant developments in astronomy, theoretical astrophysics, cosmology, high energy astrophysics and space research in the former Soviet Union. Plans for future space experiments are also highlighted.

©2004 Cambridge Scientific Publishers. All rights reserved.

Except as permitted under national laws or under the photocopy license described below, no part of this publication may be reproduced or transmitted in any form or by any means, electronic, mechanical, photocopying or otherwise, or stored in a retrieval system of any nature, without the advance written permission of the Publisher.

Ordering Information
Each volume consists of an irregular number of parts depending upon size. Issues are available individually as well as by subscription. 2003 Volume: 12.

Orders may be placed with your usual supplier or at one of the addresses shown below. Journal subscriptions are sold on a per volume basis only. Claims for nonreceipt of issues will be honored if made within three months of publication of the issue. Subscriptions are available for microform editions; details will be furnished upon request. All issues are dispatched by airmail throughout the world.

Subscription Rates
Base list subscription price per volume: EUR 182.00. * This price is available only to individuals whose library subscribes to the journal OR who warrant that the journal is for their own use and provide a home address for mailing. Orders must be sent directly to the Publisher and payment must be made by personal check or credit card. Separate rates apply to academic and corporate/government institutions. Postage and handling charges are extra.

*EUR (Euro). The Euro is the worldwide base list currency rate; payment can be made by draft drawn on EURO currency in the amount shown, or in any other currency within the eurozone at the eurodenominated rate. All other currency payments should be made using the current conversion rate set by the Publisher. Subscribers should contact their agents or the Publisher. All prices are subject to change without notice.

Orders should be placed through the Publisher at the following addresses:

Cambridge Scientific Publishers
P.O. Box 806
Cottenham
Cambridge
CB4 8RT
UK
Tel: +44 (0)1954 251283
Fax: +44 (0)1954 252517
Email: janie.wardle@cambridgescientificpublishers.com
Website: www.cambridgescientificpublishers.com

Rights and Permissions/Reprints of Individual Articles
Permission to reproduce and/or translate material contained in this journal must be obtained in writing from the Publisher. This publication and each of the articles contained herein are protected by copyright. Except as allowed under national "fair use" laws, copying is not permitted by any means or for any purpose, such as for distribution to any third party (whether by sale, loan, gift or otherwise); as agent (express or implied) of any third party; for purposes of advertising or promotion; or to create collective or derivative works. A photocopy license is available from the Publisher for institutional subscribers that need to make multiple copies of single articles for internal study or research purposes. any unauthorized reproduction, transmission or storage may result in civil or criminal liability.

Copies of individual articles may be ordered through the Publisher's document delivery service.Please contact one of the addresses listed above.

Printed in UK

April 2004

THE JETS AND SUPERCRITICAL ACCRETION DISK IN SS433

S. FABRIKA

Special Astrophysical Observatory, Russian Academy of Sciences
fabrika@sao.ru

Abstract

The review describes observations and investigations of the unique object SS433 obtained after 23 years of studying this massive binary system. The main difference between SS433 and other known X-ray binaries is the action of a constant supercritical regime for the accretion of gas onto the relativistic star (most likely a black hole), which has lead to the formation of a supercritical accretion disk and collimated relativistic jets. The properties of the jets are, to a large extent, determined by their interaction with the disk wind. The precession of the disk and jets, as well as the eclipsing in the binary system, make SS433 a unique laboratory for studies of mechanisms for the microquasar phenomenon. The review describes the main ideas and results emerging from studies of the formation of the jets and supercritical accretion disk in SS433. Essentially all photometric and spectroscopic properties of SS433 are determined by the accretion disk and its orientation, but the disk itself is not observed, being located beneath the photosphere of the dense wind. Observational manifestations of the wind and of gas flows in the system are described, as well as possible properties of the material lost by the system in its equatorial plane. Little is known about the structure of the central regions where the hot bases of the jets are located immediately above the plane of the disk. The available X-ray, ultraviolet, and optical observations paint a picture in which the bases of the jets are surrounded by cocoons of hot gas reradiating emission from the inner regions of the jet channel. Direct investigations of this channel in the supercritical accretion disk of SS433 are not possible; however, a similar object oriented face-on would likely be an extremely bright X-ray source, such as those observed in other galaxies.

Introduction

The well known and unique object SS433 was first identified in the survey of stars exhibiting Hα emission of Stephenson and Sanduleak (1977), which included 455 objects in the Galactic plane. SS433 was a variable non-thermal radio source (Feldman et al. 1978; Seaquest et al. 1978) and variable X-ray source (Marshall et al. 1978). The first slit spectra of this object (Ciatti et al. 1978; Mammano and Vittone 1978; Clark and Murdin 1978) revealed bright and variable lines whose origin was unclear. Bruce Margon and his colleagues (Margon 1979; Margon et al. 1979ab) identified these emission lines with lines of hydrogen and neutral helium shifted by tens of thousand of km/s to the red and the blue, with a pair of lines for each transition. The huge observed shifts of these lines could not arise due to Zeeman splitting (Liebert et al. 1979), and it was evident that the line shifts were due to the Doppler effect associated with rapidly moving gas. It was then elucidated that the shifted H and He I lines arise in two oppositely directed jets of gas (Fabian and Rees 1979; Milgrom 1979a; Margon et al. 1979c) that change their position in space in a periodic fashion (they "precess"), leading to "motion" of the lines in the spectrum. This was the beginning of intensive studies of SS433, as a binary system with unique properties.

The main characteristic property of SS433 that distinguishes it from other binary systems containing relativistic objects is that, in SS433, a continuous (non-transient) regime of supercritical accretion of gas onto the relativistic star is realized. In this case, a supercritical accretion disk forms, together with narrow jets of gas that propagate with the relativistic speed of 79 000 km/s from the inner regions of the disk, perpendicular to the disk plane. The second component of the system – the donor star – obviously fills its critical Roche lobe, providing a powerful and approximately continuous flow of gas into the region of the relativistic star at a rate of $\sim 10^{-4}\,M_\odot/\mathrm{yr}$. Essentially, the reason SS433 is unique among other massive X-ray binary stars (with black holes or neutron stars) can be revealed by determining the origin of the very high rate of mass transfer in this system (van den Heuvel 1981; Shklovskii 1981).

It is interesting that no direct observational evidence has been found for either an accretion disk or a "normal" or "optical" star in the SS433 system. Nevertheless, there is no doubt that these two components are present in SS433. This opinion is not only due to the accumulated experience with studies of dozens of close X-ray binary systems having neutron stars or black holes as their relativistic stellar components. There are many indirect pieces of evidence and observational manifestations of these two components, and all the main properties of SS433 can be well described using modern concepts about supercritical disk accretion, first discussed by Shakura and Sunyaev (1973).

SS433 is a close binary and a massive eclipsing system with an orbital period of 13.1 days (Crampton *et al.* 1980; Cherepashchuk 1981). Eclipses of both bodies are clearly observable in the optical and near infrared, as well as eclipses of the base of the relativistic jets in the X-ray. The source of the jets (accretion disk or object at the disk center) is appreciably brighter than the secondary (donor) star. The accretion disk of SS433 precesses, changing its orientation in space with a period of $P_{pr} = 162$ days, with the jets repeating this precessional motion. Essentially, we observe in SS433 only a dense wind outflowing from an accretion disk, and two bright regions in the central part of the disk, at the places of exit of the relativistic jets. Observationally, the star in SS433 is manifest only as an object that periodically eclipses the accretion disk and gaseous flows forming the disk, reflects the radiation of bright central regions, and perturbs the disk wind. The precession of the accretion disk cardinally changes the photometric properties (orbital light curve) and appreciably affects the spectral properties of the system. Below, we will use the term "accretion disk" to mean not only the disk itself, which must be present, but also the disk wind, and will use the term "optical" or "normal" star for the secondary, in spite of the fact that very little is known about this star.

In this review, we will describe the main properties of SS433's relativistic jets and accretion disk – the machine generating the jets – as they are currently (2002) understood, focusing primarily on the available observations and their interpretation. Spectral and photometric studies of SS433 as a binary system will also be described, since these results are required to understand the nature of the disk and jets. The bulk of observational data on SS433 were obtained in the first years of investigations of this object, during the "SS433 boom". The main ideas and models attempting to explain the behavior of SS433 were also proposed in these early years. To a large extent, these ideas obtained confirmation in subsequent observations. Therefore, previously published reviews on SS433 remain valuable. We refer the reader to these reviews, not only for information on the history of studies of SS433 – an object that has played and continues to play a fundamental role in modern astrophysics – but also because of the importance of these reviews. Margon (1984) summarized the results of five years of investigation of this object, while Cherepashchuk (1989) reviewed the results of photometric studies. Milgrom (1981), Petterson (1981), and Katz (1986) addressed models and theoretical concepts for SS433. Reviews have also been published by Clark (1985), Zwitter et al. (1989), and Vermeulen (1996). Of course, the results of new observations, especially X-ray and radio interferometric observations, as well as numerical simulations, have also made a fundamental contribution to our understanding of SS433.

Here in the Introduction, we present a very brief list of the main parameters of SS433 (many of which will be discussed in detail below), to provide a basis so that the subsequent chapters of the review can be read independently.

Parameters of SS433

SS433 is the variable star V1343 Aquilae, located at a distance of 5.0 kpc roughly in the Galactic plane ($l = 39.7°, b = -2.2°$). This is a relatively bright red star: $V = 14.0$, $(U-B) = 0.8$, $(B-V) = 2.1$, $(V-R) = 2.2$ (Goranskii et al. 1998a). A finding chart of SS433 and photometric data for the surrounding stars can be found in Leibowitz and Mendelson (1982). SS433 is strongly absorbed, $A_V \approx 8^m$, and the intrinsic luminosity of the object assuming isotropic radiation of its emission is $L_{bol} \sim 10^{40}$ erg/s (Cherepashchuk et al. 1982; Dolan et al. 1997). It is one of the brightest stars in the Galaxy, and has its spectral maximum in the ultraviolet. There is an infrared excess in the L and K bands, in which the mean brightnesses of the object are $7^m.0$ and $8^m.0$, respectively (Giles et al. 1980; Kodaira et al. 1985). This infrared excess is associated with free–free radiation by gas in the immediate vicinity of the system. The X-ray luminosity of SS433 is about $\sim 10^{36}$ erg/s (Brinkmann et al. 1991; Kotani et al. 1996; Marshall et al. 2002). The 1–10 keV X-ray emission is primarily due to hot ($\sim 10^8$ K) gas of the jets located above the photosphere of the accretion disk.

In addition to emission lines associated with both jets that shift in accordance with the precessional and nutational periods, the optical spectrum of SS433 shows very bright and variable "stationary" lines of hydrogen, He I, He II, C III, and N III, as well as other weaker Fe II emission lines (Murdin et al. 1980; Crampton and Hutchings 1981a). Together with the H I and He I emission lines, these last lines show clear P Cyg profiles at certain precessional phases. All these lines are formed both in the wind flowing from the accretion disk and in gaseous flows in the system. No lines from the normal star have been detected (Gies et al. 2002a), despite numerous attempts to do this. However the latest data show that the donor star in SS433 is an evolved A supergiant (Gies et al. 2002b).

The radiation of SS433 is very variable in all accessible wavelength ranges. In addition to sporadic variability (flares), active and quiescent states are observed. In quiescent states, optical, IR, and X-ray variability with the orbital and precessional periods is observed. In active states, which last from 30 to 90 days, the mean brightness of the object increases by approximately a factor of 1.5, and powerful flares with characteristic time scales of hours to days (Irsmambetova 1997) are observed against this enhanced background level. In addition, in active states SS433 "reddens", i.e., the gas exchange and flow of gas from the system are increased. The active periods are especially clearly visible in the radio, where long series of observations are available (Bonsignori-Facondi et al. 1986; Fiedler et al. 1987).

The Jets of SS433

The most striking phenomenon associated with SS433 is its jets. Depending on the distance from the source, or alternatively on the temperature of the jets, radiation mechanism and, accordingly, observational methods used, we can distinguish the X-ray jets ($\sim 10^{10-13}$ cm), optical ($\sim 10^{14-15}$ cm), and radio ($\gtrsim 10^{15}$ cm); the extended X-ray jets are also observed ($> 10^{17}$ cm). However, this division is somewhat arbitrary; for example, radio emission is observed virtually along the entire extent of the optical jets.

The jets are manifest in the optical spectra as "moving" emission lines of hydrogen and He I. The lines are shifted in the spectrum due to variations in the inclination of the jets to the line of sight during their precession. The jets are surprisingly narrow, and their opening angle in the region in which the hydrogen lines are emitted (whose distance from the central object corresponds to 1–3 days of flight) is 1°.0–1°.5 (Borisov and Fabrika 1987). Clouds of gas with a normal "astrophysical" temperature of $\sim 10\,000$ K move in the optical jets (Davidson and McCray 1980). A continuous source of heating is required to maintain the emission of the gas in the optical jets. The X-ray jets (Marshall et al. 2002) are short (only several hundreds of seconds of flight), and give rise to lines of highly ionized heavy elements. The jet X-rays are radiated by hot gas ($T \sim 10^8$ K), which cools as the jets propagate due to expansion and radiative cooling. The opening angle of the X-ray jets is $\approx 1°.2$. The gas in the SS433 jets moves along strictly ballistic trajectories. The flux of kinetic energy, or kinetic luminosity, of the jets is enormous: $L_k \sim 10^{39}$ erg/s (Panferov and Fabrika 1997; Marshall et al. 2002).

The well-known precessing jet pattern is observed in the radio on scales of several arcseconds (Hjellming and Johnston 1981). The radio flux from the synchrotron-radiating jets is about 1 Jy, and their luminosity is 10^{30-31} erg/s. The jets are also clearly visible on VLBI scales (Vermeulen et al. 1987), right down to the smallest available resolution of about 2 mas (Paragi et al. 1999, 2000), where the effect of self-absorption is already strong in inner regions ~ 20 AU in size. The jets of SS433 excite the radio nebula W50, which closely resembles a supernova remnant. W50 is extended in the direction of the jet precessional axis (PA$\approx 100°$) with SS433 at its centre, and the nebula stretches on both sides in this orientation to 50–70 pc. The large-scale X-ray jets extend in this same direction (Brinkmann et al. 1996), and end in optical filaments (Zealey et al. 1980).

A "kinematic model" of SS433 (Abell and Margon 1979) predicts the position of the jets in space and the positions of lines in the spectrum very well. This is a geometric model for the jet precession, and will be considered in more detail below. In spite of some instability in the precessional period, this model has been fully confirmed over long time intervals (Eikenberry *et al.* 2001). In addition to their precessional motion, the jets undergo small amplitude so-called nutational oscillations with a period of 6.28 days, which is half the synodic orbital period. The nutational nodding of the jets (accretion disk) is due to periodic tidal perturbations of the disk by the gravitational field of the donor star (Katz *et al.* 1982) or to perturbations of the accretion flow. The most successful precession scenario for SS433 is forced precession of the donor star, whose rotational axis does not coincide with the orbital axis, and the drifting or "slaved" accretion disk (Shakura 1972; van den Heuvel *et al.* 1980; Whitmire and Matese 1980; Katz 1980).

Further in the review, we will describe the results of observations of the SS433 jets, observational manifestations of the accretion disk, and SS433 as a binary system, as well as our current understanding of the physical processes associated with the jets and their formation and collimation. It is still early to state that the "mystery of SS433" has been solved; the most interesting works in connection with many questions, especially the formation of the jets and the inner structure of the central object, probably still lie ahead. However, the progress in our understanding of SS433 that has already been attained is no less surprising than SS433 itself. This object has exerted a huge influence on modern astrophysics, in our understanding of critical stages in the evolution of close binary systems and of the jets ejected from young stars, active galactic nuclei and microquasars (Mirabel and Rodrigues 1999). Microquasars appear to be the closest relatives of SS433. The main, but far from only, characteristic distinguishing SS433 from microquasars and some X-ray novae, in which episodes of supercritical accretion are possible during flares, is the presence of a constant, appreciably supercritical accretion regime onto the relativistic star. SS433 remains the only stellar-mass object in which we can directly observe an operating supercritical accretion disk, together with the ejection and propagation of the jets. Moreover, this disk (and also all gaseous flows in the system and wind from the disk and jets) is constantly turning with the precessional period and is eclipsed with the orbital period, making the system a true gift for researchers and a unique astrophysical laboratory. The two jets of SS433 should be intrinsically identical, but often appear very different. The changing orientation of the jets also presents splendid opportunities for studies of the behavior of gas moving at relativistic speeds and of the associated relativistic effects themselves.

In all, 614 papers on SS433 were listed in the ADS server (http://adsabs.harvard.edu) in January 2002; references to SS433 were made in the abstracts of an additional 199 articles (i.e., these latter papers were concerned with objects or phenomena directly related to SS433). In the year of its discovery, 1978, five articles and other communications were published; the maximum publication rate for papers on SS433 was in the following three years from 1979 to 1981, with on average 75 papers per year. Further, the number of SS433 publications steadily decreased, as was quite natural. The average number of papers for 1982–1989 was 27 per year, while the average for 1990–2001 was 14 per year. The intensity of studies of SS433 again began to rise in 1996–2000, due in part to the ability to obtain new observations in the X-ray (ASCA, ROSAT, CHANDRA) and radio (Very Long Baseline Interferometry), as well as to the discovery of a new related class of object – microquasars. The importance of this last factor is confirmed by the number of publications in which the name SS433 appears only in the abstract: the number of such articles had a broad maximum in 1981–1986 (while the frequency of direct investigations steadily decreased), but an even higher maximum in 1996–1999.

Naturally, it is not possible to reflect the results of more than 800 publications on SS433 in this review, although a substantial majority of them made appreciable contributions. Many gave new life to old topics of study or stimulated subsequent investigations, and the history of some important turning points in studies of SS433 reads almost like a detective story. The main goal of this review is to describe our current understanding of SS433.

The Optical Jets

The Jet Spectral Lines

The brightest optical lines radiated in the SS433 jets are hydrogen lines, namely $H\alpha^{\pm}$, where the "+" line forms in the receding jet and the "−" in the approaching jet. The mean equivalent width of

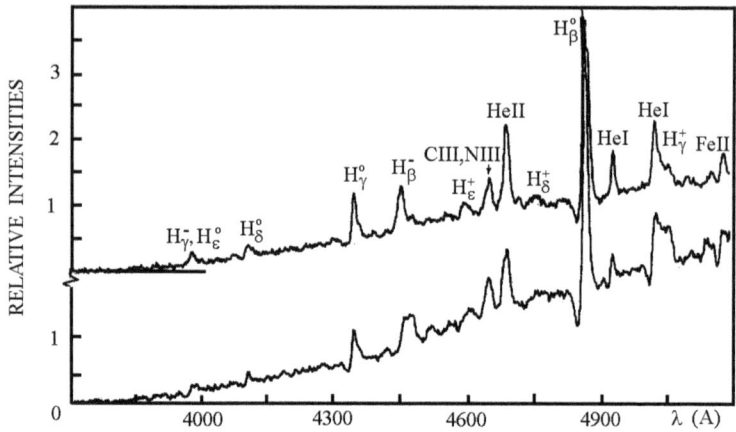

Figure 1. Blue spectra of SS433 obtained on June 1 (upper) and June 2 (lower), 1986. A small shift of the moving lines by 1 day can be seen, as well as an enhancement of the absorption components of the stationary lines.

the Hα lines is several tens of Å, and the jet lines are strongly variable. As a rule, higher-order Balmer lines emitted by the jets have not been studied in detail, since SS433 is relatively weak in the blue, and it is often difficult to distinguish the properties of individual lines in this region of the spectrum (which has an appreciably richer line structure than the red) due to blending of lines from the jets and numerous "stationary" lines. Figure 1 shows spectra of SS433 in the blue obtained on the 6-m telescope of the Special Astrophysical Observatory by Goranskii *et al.* (1987) on June 1 and 2, 1986 as part of a program of coordinated observations of SS433. Only the strongest stationary and moving lines are marked; weaker lines belong primarily to He I. Among the stationary lines, only the hydrogen, He I and Fe II lines show P Cyg profiles. As a rule, spectroscopy in the blue part of the spectrum has been used for investigations of SS433 as a binary system.

The moving H$\alpha^\pm$ lines are about an order of magnitude less intense than the stationary Hα line. The prominent moving lines include He I lines from the strongest transitions; the He I$^\pm$ lines are

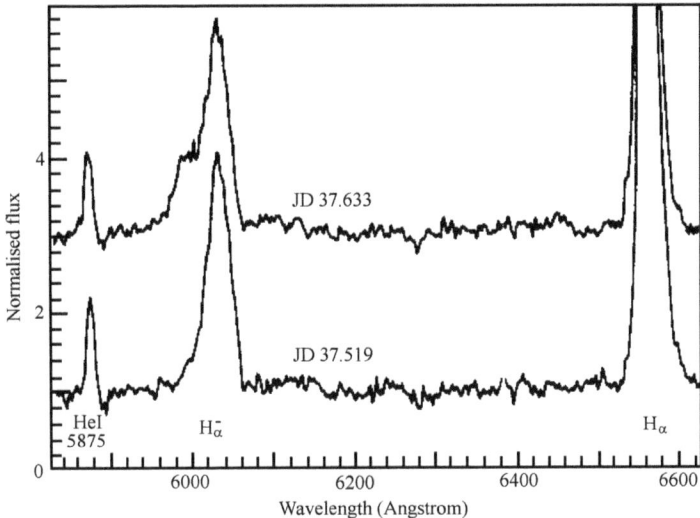

Figure 2. Fragments of two spectra of SS433 (Vermeulen et al. 1993a) that show rapid variations in the Hα profile of the approaching jet.

about an order of magnitude weaker than the H$\alpha^{\pm}$ lines, suggesting an absence of strong chemical anomalies in the SS433 gas. No He II λ4686 emission has been detected from the jets, although this is most likely due to the limited signal/noise ratio of the spectra (Vermeulen et al. 1993a). According to our estimates based on spectroscopic observations of SS433 conducted on the 6-m telescope of the Special Astrophysical Observatory, the intensity of this line in the jets does not exceed 1% of the continuum intensity.

Virtually all data on variability of the optical jets and on the geometric and kinematic structure of the jets have been derived from observations of the H$\alpha^{\pm}$ lines. Figure 2 presents fragments of two spectra of SS433 containing the Hα^{-} line (Vermeulen et al. 1993a) and the stationary Hα and He I λ5876 lines. The spectra obtained on the 1.2-m Calar Alto telescope on May 21, 1987, also during an observing campaign targetting SS433, show rapid variability of the jet lines. New volumes of gas emitting in the Hα line appeared in the jet over a time interval of three hours.

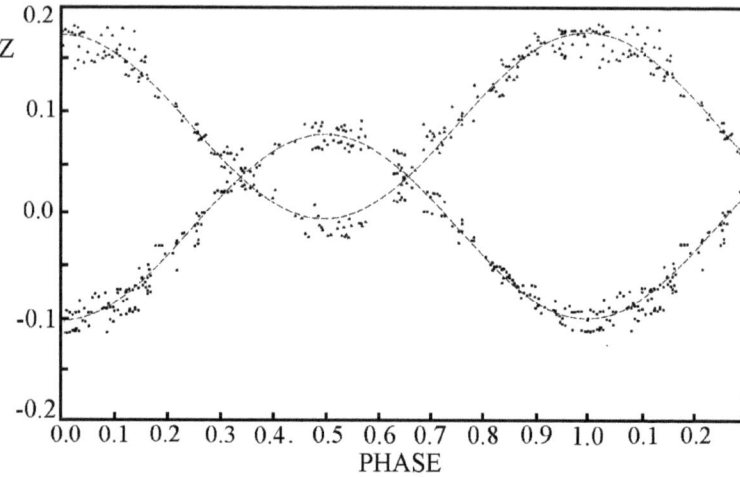

Figure 3. Precessional curves of the radial velocities of the shifted lines from the approaching (lower curve) and receding (upper curve) jets, derived from spectroscopic data obtained during the first two years in which SS433 was studied (Ciatti *et al.* 1981). The scatter of the data about the curves is due primarily to the nutational variability.

Kinematic Model and Precession of the Jets

Variations in the radial velocities of the jets with the precessional phase derived from the $H\alpha^{\pm}$ lines (Ciatti *et al.* 1981) using data for the first two years of SS433 studies are shown in Fig. 3. The mean radial-velocity curves are shown, with the scatter of the data about these curves being due to the nutational variability. During the precessional period, the jets lie in the plane of the sky twice (the radial velocities of the two jets coincide), two crossovers of the moving lines are observed, and, accordingly, the jet lines also move away from each other twice. The time of maximum separation of the lines to the blue and red corresponds to the minimum inclination of the jets and the axis of the accretion disk to the line of sight, which occurs at precessional phase $\psi = 0$, also called the T_3 moment. The two crossovers are usually denoted times T_1 and T_2, and their precessional phases are 0.34 and 0.66. It is obvious that the

THE JETS IN SS433

phases for the extrema and crossovers of the radial-velocity curves (Fig. 3) are determined only by the orientation of SS433 relative to the observer and not by any physical processes occurring in the system. This, generally speaking, trivial fact nevertheless is sometimes forgotten in interpretations of the complex phenomena observed in SS433.

During the precessional cycle, the lines of the two jets change places, so that the jet that recedes from us during most of the precessional period is denoted with a "+" and the opposite jet is denoted with a "−". At the times $T_{1,2}$, the radial velocities of the jet lines coincide, but are not equal to zero. This is due to the well known transverse Doppler effect, or time dilation, which is clearly observed in this way only in SS433 (among macroscopic objects). The Doppler shifts of the spectral lines are described by the well known formula

$$\lambda = \lambda_0 \gamma (1 - \beta \cos \eta),$$

where λ and λ_0 are the shifted and laboratory wavelengths, η is the angle between the jet and the line of sight,

$$\gamma = \frac{1}{\sqrt{1-\beta^2}}$$

is the Lorentz factor and the jet velocity V_j is expressed in units of the speed of light $\beta = V_j/c$. At the times of the crossovers, the jet radial velocities are $V_r^\pm/c = \gamma - 1$. Thus, the velocity of propagation of the jets can be directly measured in SS433, and, consequently, the geometric parameters of the jets, inclination of the system, and distance to the object (from radio images of the precessing jets) can also be determined.

The behavior of the moving lines of SS433 is described well by the kinematic model for the precessing jets (Abell and Margon 1979). Figure 4 shows a geometric schematic of the jet precession. We adopt the following notation. The angle between the jets and the precessional axis (the precession angle) is θ, the angle between the precessional axis (orbital axis) and the line of sight i, the precessional period P_{pr}, and the precessional phase ψ. The angle between the approaching jet and the line of sight is η, $\cos \eta = \sin i \cos \theta + \sin \theta \cos i \cos \psi$, and this angle is minimum when $\psi = 0$. The radial velocities of the

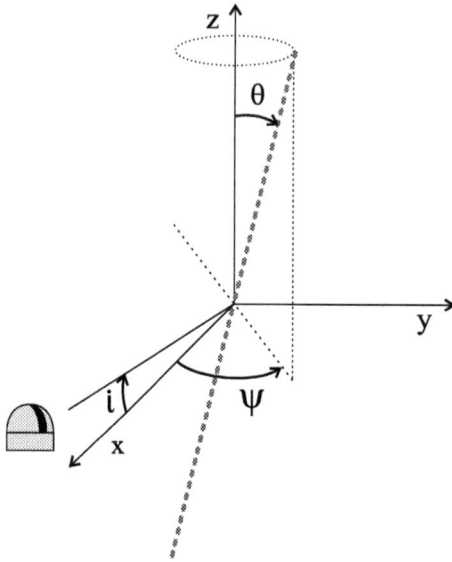

Figure 4. Geometrical schematic of the jet precession. The precessional axis is Z and the line of sight is in the XOZ plane.

two jets $V_r^\pm/c = z^\pm$, or equivalently the positions of the lines in the spectrum, can be computed using the formula

$$1 + z^\pm = \gamma(1 \pm \beta \sin\theta \sin i \cos\psi \pm \beta \cos\theta \cos i),$$

where the "+" and "−" signs correspond to the receding and approaching jets. This kinematic model was tested and refined after four years (Anderson et al. 1983), ten years (Margon and Anderson 1989) and twenty years (Eikenberry et al. 2001) of spectroscopic studies of the SS433 jets. The analysis of Eikenberry et al. (2001) is based on 433 values for z^+ and 482 values for z^-. To avoid uncertainty associated with the 6.3-day period of the jet nutation, the data were smoothed with a broader time filter. Thus, the kinematic model can be used to study only the precessional motion and possible long-term deviations and secular variations in the precessional clock of SS433. The mean values of the precessional parameters of SS433 were determined to high accuracy by Eikenberry et al. (2001): $\beta = 0.2647 \pm 0.0008$, $\theta = 20°.92 \pm 0°.08$, $i = 78°.05 \pm 0°.05$, $P_{pr} = 162^d.375 \pm 0^d.011$; the date corresponding to the T_1 moment is $JD = 2443563^d.23 \pm 0^d.011$. The epoch T_3 of the maximum separation of the lines in the spectrum, or the precessional phase $\psi = 0$, occurs at $JD = 2443507^d.47$. These values are the result of finding the optimal five-parameter precessional model. The actual kinematic parameters could differ slightly from these values; for example, a simple average of the jet velocity yields $\beta = 0.254 \pm 0.0011$, which is 3 200 km/s lower than the velocity given by the kinematic model.

Note that we should not smooth the nutational variability to derive the real precessional trajectory. The nutational deviations are due to tidal perturbations of the accretion disk by the gravitational field of the donor star (Katz et al. 1982; Collins and Newsom 1986). These perturbations lead to a periodic decrease in the angle between the plane of the disk and the plane of the orbit. Therefore, the real surface of the precession cone passes closer to the outer extrema of the nutational trajectory of the jets. Taking this effect into account leads to a slight increase in the precession angle θ by an amount equal to the nutational angle, $\approx 3°$.

The precessional period has been stable over a long time interval of about 20 years (Eikenberry et al. 2001), $\dot{P}_{pr} < 5 \cdot 10^{-5}$, in spite of numerous reports of variations of this period in the first few years of studies of the object (Anderson et al. 1983). This behavior is associated with real instability of the precessional cycle on time scales of weeks to months, which cannot be distinguished at any precessional phases. This instability leads to the appearance of real (but random) variational trends of the period over times of several hundred days. Instability in the precession has also been detected in optical photometric measurements (Goranskii et al. 1998b). The mean brightness of the system varies with the precessional phase by roughly $0^m.5$ (Kemp et al. 1986; Gladyshev et al. 1987); SS433 becomes brighter at the time T_3, when the disk is maximally turned toward the observer (the angle between the disk axis and the line of sight is 57°). Optical photometry is used as an independent method for studying the precessional clock.

In a model of driven precession (for example, of the precession of the rotational axis of the normal star), the precessional and orbital periods are directly related. Variations in the orbital period of SS433 based on the times of eclipses (Fabrika et al. 1990; Goranskii et al. 1998b) in O–C diagrams also show instability of about the same amplitude and at about the same times as the instability in the precessional period. Small-scale instability in the precessional and orbital clocks could be associated with variations in the rate of mass transfer between the components in the active and passive states of SS433. Like the precessional period, the orbital period is also stable over long time scales (Goranskii et al. 1998b).

The instability of the precessional cycle (Margon and Anderson 1989; Baykal et al. 1993; Eikenberry et al. 2001) resembles random deviations of the precessional phase from the computed ephemerides with amplitudes to $\Delta\psi \approx 0.1$ (7–15 days), and is probably associated with both real variations in the phase and variations in the inclination and speed of the jets. None of the parameters θ, β or P_{pr} by itself can explain the observed precessional "noise". The statistical behavior of the deviations is well described as white noise in frequency, or as a random walk of the phase of the precessional period (Baykal et al. 1993). The precessional noise of SS433 is quantitatively similar to the noise in the 35-day (precessional) period of the X-ray source Her X-1.

Almost no observations of the moving lines have been made over the past ten years (at least the results have not been published); however, investigating the origin of these instabilities requires series of spectral monitoring observations. For example, it would be useful to compare the times when the instabilities appear with periods of activity of SS433. In their analysis of the periodicity in the anticorrelated shifts of the lines of both jets in the spectrum, Frasca et al. (1984) detected about ten harmonics, including harmonics corresponding to periods of 80, 155 and 1500 days. No periodicity in the absolute velocity of motion of the jets was found. Such analyses should be continued and refined using additional data.

How the Moving Lines Move

The SS433 jets are strictly antisymmetrical, and, as a rule, the profiles of lines radiating in the opposite jets show mirror symmetry. The arrival times for the signals from the two jets should differ only

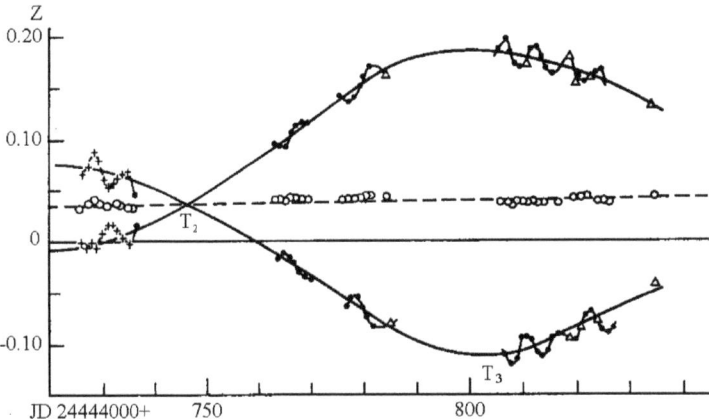

Figure 5. Fragments of the radial-velocity curves for moving lines according to the data of Kopylov *et al.* (1987), constructed using the strongest components of the H$\alpha^\pm$ lines. The nutational behaviour of the jets is completely antisymmetric.

slightly. The region of maximum brightness in the hydrogen lines is separated from the central source by approximately one day of flight of the gas in the jets. Even when the inclination of the jets to the line of sight is maximum (at the time T_3), the time delay for the radiation of the receding jet is about 0.2–0.25 day. Therefore, rapid reconstruction of the jet structure such as that shown in Fig. 2, or even more rapid line profile variations (on time scales of less than an hour; Kopylov *et al.* 1986), are often observed only in one (as a rule, the closer) jet. However, overall, the differing arrival times of the signals in no way disrupts the observed symmetry of the jets.

In addition to the regular nutational motion of the two jets (Fig. 5), short-term (time scales of several days) disruption or jitter of the jets is sometimes observed. The amplitude of the jitter reaches 3 000–5 000 km/s, which is equivalent to variations in the jet inclination by 2°.5–4°. The jitter amplitude is comparable to (or slightly higher than) the amplitude of the nutational motion. The origin of the jitter is probably closely related to the nature of the nutational shifts of the flows of accretion gas, as well as to the conditions for the development of a certain disk inclination (or, more precisely, to the disruption of these conditions), and the time for the passage of matter through the disk. For example, in a slaved-disk model, the instantaneous inclination of the disk depends on the inclination of the star, the orbital phase (periodic gravitational moment perturbing the disk), specific geometry for heating of the star by the bright source (shadowing of part of the stellar surface by the edge of the disk or by clouds of gas), and the accretion rate (state of activity).

The jet lines sometimes "disappear" for up to several days, after which they appear in the expected position, in accordance with the ephemerides (Kopylov et al. 1985; Vermeulen et al. 1993a). It is not ruled out that such "switching off" of the engine is somehow related to active periods; in both papers cited above, the switching off of the jet emission coincided with powerful photometric flares of the object. It remains completely unclear whether these disappearances are associated with the cessation of jet activity, or possibly with the disruption of the jet-collimation mechanism and the onset of thermal instabilities that result in the formation of clouds of cool gas in the jets. A detailed analysis of the times of disappearance of the jets could shed light on the mechanism for collimation and acceleration of the jets.

The answer to the question of "how the moving lines move" (Grandi and Stone 1982) is now well understood. The jet gas travels along strictly ballistic trajectories (straight lines), along which it was ejected from the source. The source – central region of the accretion disk – participates in the continuous precessional and nutational motion of the system. The ejection of the gas in the jets is modulated, and occurs in clumps, with on average one to three clumps ejected per day. These clumps are sometimes called "bullets", and appear in the spectra (the "young" jet) as fragments of line profiles or individual lines that have the invariable position in the spectrum. The radiation of these bullets is already appreciably weaker after a day, and they remain as weakening "remnants" (the "old" jet) in the spectrum for up to four, or sometimes even six, days after their appearance. The more rapidly the lines are shifted through the spectrum (i.e., the more rapidly the angle between the jets and the line of sight changes), the less energy is accumulated at the given wavelength. Therefore, the moving lines and their numerous components are most clearly visible at the phases of extrema of the precessional and nutational periods, when the inclination of the jets to the line of sight varies slowly. In terms of the precessional motion, these are phases 0.0 (the time T_3) and 0.5.

The angular velocity of the nutational motion is fairly high, so that a line will be spread over virtually the entire spectrum between the extrema of the nutational curve, and, on the contrary, will be appreciably strengthened at the extrema. This is a geometrical projection effect (Borisov and Fabrika 1987). Thus, the profiles of the moving lines depend on the nutational and precessional phases. As a rule, there is one bright component (FWHM = 1 000–1 500 km/s) formed by the effect of projection, as well as several weaker secondary components within an interval of several thousands of km/s. The secondary components are remnants ("tracks") of either the previous bright component or the largest bullets.

Many interesting data on the SS433 jets were obtained in coordinated observations in May/June 1987 (Vermeulen et al. 1993a), during which about 200 spectra were obtained over 20 days at various observatories around the world. In addition, the campaign included radio interferometric observations of the jets (Vermeulen et al. 1993b), radio monitoring (Vermeulen et al. 1993c), optical photometry (Aslanov et al. 1993) and X-ray observations (Kawai et al. 1989). Figure 6 shows the result of spectral observations of the moving $H\alpha^\pm$ lines. The non-stationary injection of new "bullets" of material into the jets against the background of the regular precessional and nutational motions is clearly visible.

Geometric and Kinematic Parameters of the Jets

Although the bullets appear in the jets relatively suddenly, over a few hours, the weakening of their emission lasts over several days, and can be studied in detail. Kopylov et al. (1987) and Vermeulen et al. (1993a) present light curves for individual clumps of material. Borisov and Fabrika (1987) used the data of Kopylov et al. (1987) to derive the brightness profile along the jets in $H\alpha$ emission at the phase of weakening of the emission ($R \geq R_m$):

$$F(H\alpha) \propto \exp(-(R-R_m)/R_f),$$

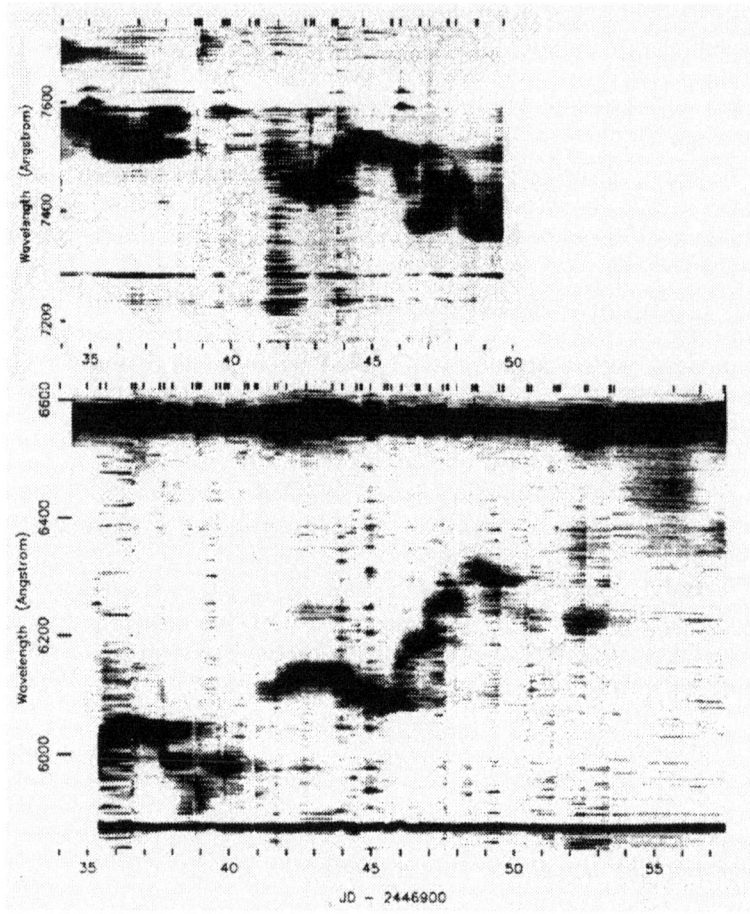

Figure 6. Grey-scale representation of the behaviour of the moving $H\alpha^+$ (upper) and $H\alpha^-$ (lower) lines based on the coordinated observations reported by Vermeulen et al. (1993a). The vertical axes plot the wavelength and the horizontal axes the observation epochs. The darkest places correspond to the brightest parts of the line profiles. The stationary $H\alpha$ and He I $\lambda5876$ lines (lower) and He I $\lambda7281$ line (upper) are visible as horizontal bands. The vertical bars in the upper parts of both panels mark the positions of individual observations.

where the maximum radiation occurs in the jet at a distance of $R_m \approx 4 \cdot 10^{14}$ cm from the source (0.6 days of flight), and the characteristic scale for the decay of the radiation is $R_f = (6.7\pm0.5) \cdot 10^{14}$ cm (1.0 ± 0.07 days of flight). This behavior is obeyed for variations of $F(H\alpha)$ by more than 1.5 orders of magnitude, and the radiation of the remnants of the moving lines can be reliably traced for four days.

Modeling of the profiles of the moving $H\alpha^-$ line (see also Panferov and Fabrika 1993) showed that, at the phase of ignition of the emission, i.e., when $R < R_m$, the brightness profile of the jet can be described by the relation $F(H\alpha) \propto \exp((R - R_m)/R_{in})$, where $R_{in} \leq 1.7 \cdot 10^{14}$ cm (0.25 days of flight). New clumps of gas at the base of the jets cool and begin to intensely radiate hydrogen line emission very rapidly, after only a few hours. Vermeulen et al. (1993a) found that the total ignition time for new clumps of gas in the jets was 6–10 hours. Thus, the brightness profile of the SS433 jets has been firmly established.

The geometrical and kinematical parameters of the jets were determined by Borisov and Fabrika (1987) based on modeling of the profile of the moving $H\alpha^-$ line (Fig. 7). Their model jet underwent precessional and nutational motion and was filled with clouds of gas at its base, which were radially distributed across the jet in accordance with a normal distribution with standard deviation $\theta_j/2$. The gas moved along ballistic trajectories at constant speed. Kopylov et al. (1986) suspected that the gas in the jets was decelerated by $\Delta V_j/V_j \leq 10^{-2}$, corresponding to a shift of or no more than 10 Å over several days of flight. This effect is weak, and even if it does exist, it does not significantly affect the structure of the computed line profiles. The opening angle of the jets was found to be $1°\!.0 < \theta_j < 1°\!.4$, with both the natural opening angle of the jet θ_j and the parameters of the nutational trajectory making significant contributions to the observed width of the moving line. The nutation angle is $\theta_n = 2°\!.8 \pm 0°\!.3$. The typical structure of the line profiles suggests that the number of clouds (gas clumps) in the jet is $\sim 10^3$–10^4, or that the time to form one such cloud is ~ 100 s. However, the rate at which gas enters the jet is also variable on time scales of about $0^d\!.3$–$0^d\!.5$ (Borisov and Fabrika 1987; Vermeulen et al. 1993a). This sporadic activity (bullets) creates large-scale structure in the line profiles.

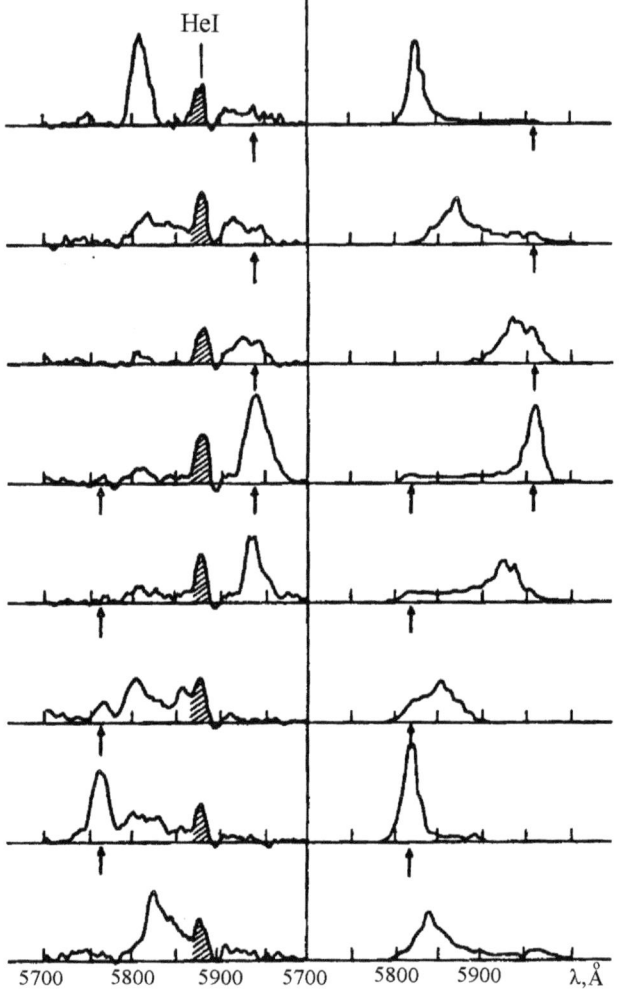

Figure 7. Observed (left) and modelled (right) profiles of the $H\alpha^-$ moving line, with the date increasing upward. The observational data were obtained on June 21–28, 1981 by Kopylov et al. (1986, 1987). The stationary $He\,I\,\lambda5876$ line is shaded. The arrows mark evolving components at wavelengths corresponding to extrema of the nutational radial-velocity curve.

Below, when describing the physical state of the gas in the optical jets, we discuss evidence that the gas in the jets is located in still smaller clumps (in clouds or cloudlets) with sizes $l \sim 10^8$ cm, which form as a result of thermal instabilities as the gas cools. Thus, we can speak of a hierarchical structure for the jets: (i) small cloudlets, (ii) larger clouds that form on time scales of ~ 100 s, (iii) large-scale clumps of gas that form on time scales of $0\overset{d}{.}3$–$0\overset{d}{.}5$. As a rule, the term "bullets" adopted in early studies of SS433 referred to bright emission components of lines, which formed due to the effect of projection. In the light of our current more detailed understanding of the jet structure, this term may appear somewhat inappropriate, and we ascribe it here to these last large-scale inhomogeneities. The time scale of ~ 100 s coincides with the time that the jet propagates inside the funnel of the accretion disk. This can be considered evidence that the jets are collimated there, and that the observed inhomogeneities form due to thermal or hydrodynamical instabilities during the time that the jets move inside the funnel. The time scale of about 0.3–0.5 days could be comparable to the characteristic time scale for instabilities in the outer parts of the accretion disk associated with the formation of spiral shock waves, i.e., with processes that modulate the rate of gas transfer in the central part of the accretion disk.

The Radio Jets and W50

The Uniqueness of the SS433 Radio Jets

SS433 is a very bright radio star, whose central radio source radiates at a level of about 1 Jy at centimeter wavelengths. Virtually all of the emission of SS433 is non-thermal synchrotron radiation from the jets. The structure of the precessing jets was directly visible in the first maps obtained with the VLA (Hjellming and Johnston 1981). In spite of the fact that the overall dimensions of the radio jets are a factor of a hundred larger than the Hα jets, the phase of the radio-jet precession (orientation) is in good agreement with the kinematic model. Clearly, the gas in which the radiating electrons are generated moves along the same ballistic trajectories as the Hα clouds, and is directly related to them. This makes it possible to measure the distance to SS433 (5.0 kpc) with an accuracy that is unprecedented in astronomy, about 5–10%. The maximum of the radio emission occurs at a distance of $\sim 10^{15}$ cm from the source (Hjellming and Johnston 1981; Romney et al. 1987; Vermeulen et al. 1993b), from the same place where the maximum optical jet line emission originates. The brightness of the radio jets gradually decreases to a distance of $\sim 10^{17}$ cm, beyond which the jets are not visible until a distance of $\sim 10^{20}$ cm. Here, the jets are decelerated and the large-scale X-ray jets are observed together with an increased intensity of the radio emission (Brinkmann et al. 1996). This forms the so-called "ears" of the W50 radio nebula, which contain regions of X-ray and optical emission.

Figure 8 shows radio images of SS433 on various angular scales taken from Paragi et al. (2000). The figure shows (a) the W50 radio nebula depicted as a mosaic of VLA images at 1.4 GHz (Dubner et al. 1998), (b) an image of the precessing jets on scales $\sim 10^{17}$ cm obtained with the MERLIN interferometer and a global VLBI array (EVN+VLBA+Y1) at 1.6 GHz, and (c) images of the inner jets obtained using the EVN, VLBA, MERLIN, and the VLA at 1.6 GHz. In this last panel, both a region of brightening of the radio emission at a distance of about 50 mas along the jets and a weak radio structure perpendicular to the jets are visible (see below).

The radio spectrum of SS433 (Seaquist et al. 1980) is a typical synchrotron spectrum, with a spectral index $\alpha \approx -0.6$ ($S_\nu \propto \nu^\alpha$) at frequencies of 0.3–22.5 GHz. A turnover or flattening of the spectrum is observed at $\nu \sim 0.3$ GHz, which could be due to synchrotron self absorption. However, it is no less likely that it is the result of free–free absorption of the radio emission by gas of the accretion-disk wind.

The radio emission is radiated by relativistic electrons that travel with the jets and are continuously generated in them. The jets of SS433 are "heavy", being composed of a $e^- p^+$ plasma, and propagate at a relatively low speed (compared to known microquasars), $V_j = 0.26c$. It is of fundamental importance here that the SS433 jets are made up of dense clouds of gas that are capable of propagating to appreciable distances without appreciable deceleration. Therefore, these jets can be observed spectrally in the optical and X-ray, and in images in the radio and X-ray. This is what determines the uniqueness of the SS433 jets; more precisely, SS433 represents an evident

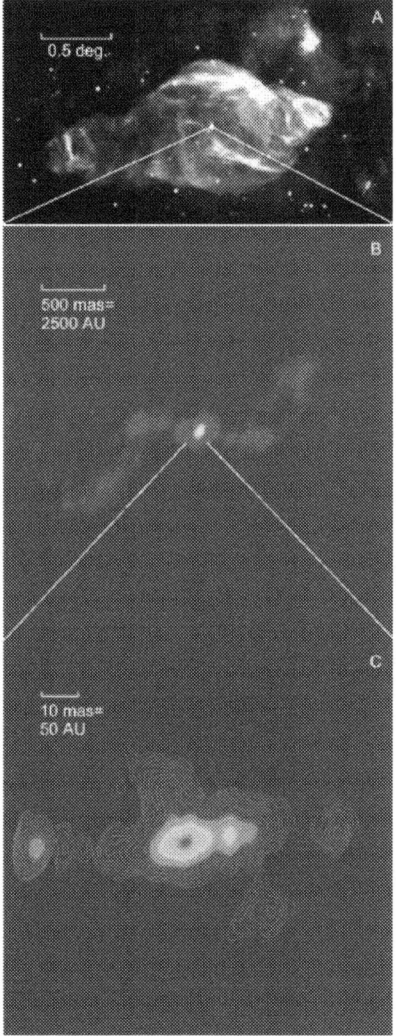

Figure 8. Radio images of SS433 on various angular scales (Paragi *et al.* 2000). (a) The W50 nebulosity (Dubner *et al.* 1998) at 1.4 GHz, (b) the precessing radio jets, (c) the inner jets, where the radio brightening zone about 50 mas along the jet and the weak radio emission approximately perpendicular to the jets are both visible.

and undoubted example of heavy jets. The relativistic particles are most likely accelerated in shock waves during interactions between the jets and gas flowing from the accretion disk. It is interesting that such interactions between the jets and the slow disk wind must also be invoked to explain the longevity of the optical radiation of the jets, as well as the six-day variations in the radio flux of SS433 (see below).

The jet radio emission is significantly linearly polarized at 10–20% (Seaquist 1981; Niell et al. 1981); the polarization is variable and oriented along the "instantaneous" direction of the jets. Circular polarization of the radio emission has also recently been discovered (Fender et al. 2000). The degree of circular polarization is 0.3–0.6% at 1–9 GHz, and its spectrum is $V \propto \nu^{-0.9\pm0.1}$. Circular polarization can in principle arise directly as a result of the synchrotron radiation of the relativistic electrons in the jets. The inferred magnetic field in this case is ~ 50 mG (Fender et al. 2000), i.e. roughly the same order of magnitude as is required to explain the low-frequency turnover of the spectrum (Seaquist 1981). Fender et al. (2000) proposed that the observed circular polarization arises via the conversion of linear to circular polarization during the propagation of the radiation through a plasma with elliptical (or linear) transmission modes. In this case, the degree of circular polarization could be as high as 10%.

To understand the processes occurring in the jets, it is also important that a variable linear polarization in the UV has also been detected (Dolan et al. 1997). The radiation is polarized to 10–15% near 2800 Å, with the orientation of the polarization coinciding with the jet direction, as in the radio. The origin of this polarization is not clear (see the section "The Supercritical Accretion Disk and the Components from Photometric Data" below). If it is associated with the jets, it arises at a distance from the source no larger than the dimensions of the binary system, most likely in the places where the jets appear above the wind photosphere. The maximum emission of SS433 is observed precisely in the UV, and the source of this radiation is the accretion disk or a region directly above the disk that is eclipsed by the secondary component in the system.

Radio Variability

Long-term monitoring of SS433 in the radio (Johnston et al. 1981, 1984; Bonsignori-Facondi et al. 1986; Fiedler et al. 1987; Bursov and

Trushkin 1995) provides an excellent demonstration of the active and quiescent states of the object. The low-frequency radio variations lag the variations at higher frequencies by on average several days. The radio jets observed with VLBI are present in both active and quiescent periods (Romney et al. 1987; Fejes et al. 1988; Vermeulen et al. 1993b). In quiescent phases, the radio flux of SS433 shows only moderate variations, up to 10%, but powerful, often overlapping flares are observed during active phases. The duration of active states ranges from 30 to 90 days, with the duration of individual flares being from one to several days.

There is no observed dependence of the times of "switching on" (i. e. transitions to active states) of the object on the orbital or precessional phases. Likewise, there are no detected variations of the radio flux itself with the orbital or precessional periods. However, the radio flux does vary with the 6-day nutational period. Johnston et al. (1981) detected such variability during an active period, when the radio emission was dominated by flares.

Band and Grindlay (1984) suggested that, in the slaved-disk model (in which the rotational axis of the optical star is inclined to the orbital axis), flares should occur twice per orbital period in a reference frame rotating with the equatorial plane of the star; i.e., with a period of 6.06 days, due to variations of the volume of the critical Roche surface with this period. The decrease in the effective Roche surface should occur when the relativistic star is in the node line; i.e., in the equatorial plane of the donor star.

In the same slaved-disk model, twice during each orbital period, there should be a perturbation of the accretion disk due to the gravitational torque from the donor star acting on the outer edge of the disk. The maximum perturbation of the disk arises when the donor is perpendicular to the node line, here it is a line of intersection of the disk and orbital planes. This well-known nodding mechanism for producing the motions of the accretion disk (Katz et al. 1982) provides the most natural explanation for the nutational motions of the jets. Both mechanisms (Band and Grindlay 1984; Katz et al. 1982) can in principle modulate the mass transfer or the jet activity in SS433. We will return to these mechanisms in the next section, in connection with our description of flares in SS433.

Here, it is important to note that, if the six-day variations in the radio flux (and the optical flux; see the section "The Supercritical Accretion Disk and the Components from Photometric Data") are associated with a restructuring of the accretion structures or with real variations in the activity of the object, we expect the variability to have the synodic period of 6.06 days:

$$f_{6.06} = 2f_{orb} + 2f_{pr},$$

however, if this variability is associated with geometrical or projection effects (as are the nutational motions of the jets), we expect the variability to have a period of 6.28 days:

$$f_{6.28} = 2f_{orb} + f_{pr}.$$

Trushkin et al. (2001) detected variations with a 6-day period during a quiescent phase of the radio emission. These variations may be associated with the nutational nodding of the jets and the corresponding variations in the relativistic beaming of the radiation. Although the amplitude of the variation of the jet inclination to the line of sight due to the precessional rotation is appreciably higher than that due to nutation ($\pm 20°$ as opposed to $\pm 3°$), no precessional variations have been detected, probably due to the strong sporadic variability (the active states) of SS433 on time scales comparable to the precessional period. This is consistent with the fact that the 6-day variability has been detected only in relatively short datasets or during the quiescent state of the object. However, note that the conditions for interaction of the jets and the slow disk wind, and consequently also the conditions for the generation of relativistic electrons, should also vary with the 6-day period (Panferov and Fabrika 1997). The effect of the interaction is maximum at phases of the nutational period when the nutational and precessional shifts of the jets add. This mechanism for the 6-day modulations operates in the immediate vicinity of the source, at $\sim 10^{14}$ cm, since this is the distance travelled by the wind during six days of revolution of the jets. In the case of the 6–day variations are connected with the jet–wind interaction conditions we expect variability of the radio flux with the synodic period of 6.06 days.

THE JETS IN SS433 29

Thus, the nature of the six-day variations in the radio flux in the active and quiescent states of SS433 remains at present unclear, although there is a good basis to hope for progress in this area. This will require accurate measurements of the period of the variations (6.06 or 6.28 days) in both the active and the quiescent states separately, and a comparison of the phases of these variations with the known ephemerides of the orbital and nutational periods. Possible origins of the variability include both periodic variations in the accretion-disk structure (Band and Grindlay 1984; Katz et al. 1982) and purely geometrical effects. Depending on the mechanism that is operating, we can expect various times delays of this variability relative to the orbital or nutational photometric variability.

Flares

During flares, the structure of the inner radio jets can undergo dramatic variations, and one-sided jets are sometimes observed (Romney et al. 1987). It is likely that the specific radio structure of flares depends not only on the asymmetry of the ejection, but also on the interaction of the jets with the surrounding gas from the disk wind of SS433 and the absorption of radio emission in this gas. The radio spectrum changes appreciably during flares. As a rule, it becomes flatter, and the low-frequency turnover is shifted to 2–3 GHz. Detailed analyses of individual flares (Seaquist et al. 1982; Band and Grindlay 1986; Vermeulen et al. 1993c) shows that there are at least two types of flares. In the first type of flare, the flux at the flare maximum is approximately the same at all frequencies, with the maximum first being reached at higher frequencies and then gradually shifting toward lower frequencies. The other type of flare is more complex, and they have a threshold frequency near 1–3 GHz, below which their behaviour is similar to flares of the first type. Above this frequency, the maximum emission is reached simultaneously at all frequencies, but the flux at the maximum decreases with increasing frequency. It is possible that the first type of flare is encountered either during the quiescent state of the object or at the beginning of the active state. Neither the first nor the second type of radio flare are in agreement (Vermeulen et al. 1993c) with the standard model for a single injection of relativistic electrons, followed by adiabatic expansion of the electron cloud (Shklovskii 1960; van der Laan 1966). The observed kinetics of the flares requires a continuous generation of relativistic particles. This is in good agreement with the idea that the interaction between the jets and wind is sharply increased during the flares.

It is interesting that the existence of two types of optical flares has also been discussed (Kopylov et al. 1985; Goranskii et al. 1998a). The first type of flares are "white" in terms of their $UBVR$ colours and have large amplitudes, and the second are red flares. When an active period begins, SS433 reddens (Irsmambetova 1997; Goranskii et al. 1998ab), and a more powerful circumstellar gas envelope develops at the activity maximum. In spite of the numerous observations of SS433 that have been obtained, studies of the development of the flares in the radio or optical that include spectroscopy are insufficient to enable firm conclusions. The flares in SS433 are the result of a perturbation in the jet activity, and spectral monitoring of the jets will be necessary if we wish to understand their nature. Here, we present the main regularities in the development of flares in SS433 based on two sets of continuous, prolonged observations: the optical spectroscopy and photometry of Kopylov et al. (1985) and the radio monitoring and interferometry and optical spectroscopy and photometry of Vermeulen et al. (1993abc). Some of these regularities may not be confirmed by future observations, but they are clearly visible in these data, and can be extremely useful for improving our understanding of the flare mechanism. In the quiescent state of the object, on the day on which an optical flare (of the first type) develops or slightly earlier, 1–2 days before the flare is detected, the optical jets "disappear". More precisely, the intensity of the jet lines decreases substantially, and the positions of the lines deviate sharply from the locations predicted by the ephemerides. A dip in the radio emission is also observed at this time. Further, an optical and radio flare is observed, and the jet lines appear at their expected locations with enhanced intensity. A new radio "blob" appears in the VLBI jet at the time of the radio flare.

This behavior suggests scenarios in which the initial flare occurs due to a perturbation (for some reason) of the inclination of the jets and the resulting conflict between the jets and the disk wind in the immediate vicinity of the source. It is possible that the interaction between the jets and wind occurs directly in the funnel of the supercritical disk, if the jets deviate by an angle that is larger than half the funnel opening angle. The amplitude of the nutational nodding of the jets is $\approx 3°$, the amplitude of the jet jitter is $2°\!\!.5\text{–}4°$, and the amplitude of jet deviations that could lead to flares is probably higher still. This flare mechanism – large deviations of the jets and their interaction with the walls of the funnel – places fairly rigid constraints on the time scale for variations in the jet inclination. If the size of the wind photosphere (see below) is $\sim 10^{12}$ cm and the wind speed is $\sim 10^8$ cm/s, the funnel will be refreshed over a time scale of only several hours. After the initial flare, there could be perturbations of the outer parts of the accretion disk, the atmosphere of the donor star and the surrounding gas, leading to an active state of the object accompanied by multiple flares.

The mass-exchange rate in SS433 is appreciably supercritical, $\dot{M} \sim 10^{-4}\,M_\odot/\text{yr}$ (Shklovskii 1981; van den Heuvel 1981), so that flares are most likely not associated with changes in the accretion rate, but instead with perturbations of the accretion disk accompanied by perturbations in the jet inclination. Possible mechanisms for such perturbations are described above (Band and Grindlay 1984; Katz et al. 1982). In addition, it is possible that SS433 has a small orbital eccentricity. For example, purely due to the effect of non-circularity of the orbit, the variations in the volume of the critical Roche lobe can reach 2% once per each orbital period (Band and Grindlay 1984), even if the eccentricity is as small as $e \approx 0.01$. When the components pass through periastron and the node line of the equator of the star (Band and Grindlay 1984) simultaneously coincides with the apsidal line (or, alternatively, the node line of the accretion disk is perpendicular to the apsidal line; Katz et al. 1982), rather strong perturbations of the Roche lobe (or accretion disk) are possible. Analysis of the intervals between eclipses of the disk and star (Min I and Min II) shows that the orbit of SS433 is close to circular, $e < 0.05$ (Fabrika et al. 1990). It was recently found that the brightest optical flares in SS433 occur in a distinct and very narrow interval of orbital phases (Fabrika and Irsmambetova 2002). This can be taken as evidence of a modest eccentricity in the SS433 orbit, since a non-circular orbit provides the only explanation for flares that occur in distinct phases of the orbital period.

It follows directly from the observations that the region in which the radio flares arise is ~ 100 AU, or 20 mas, from the source (Vermeulen 1993bc). The time delay of the radio flares relative to the optical flares is from several hours to several days. It is likely that different types of flares manifest the connection between optical and radio activity in different ways. Accordingly, not all radio flares are accompanied by an optical flare. There are relatively few data on X-ray flares (Grindlay et al. 1984; Kotani et al. 2002), since prolonged monitoring is required to analyse the flares. The X-ray emission of SS433 originates primarily in the jets and the surrounding gas, in the immediate vicinity of the source, at a distance of $\sim 10^{11-13}$ cm. We will describe the X-ray data and parameters of the X-ray jets in more detail in the next section.

Radio Brightening Zone

There is a radio "brightening zone" in the inner radio jets of SS433, observed in VLBI images at distances of ≈ 50 mas from the central object (Romney et al. 1987; Vermeulen et al. 1987, 1993b). After the appearance of new radio blobs at the center, their strength appreciably weakens as they move outward, but grows again as they pass through this brightening zone, sometimes to values even exceeding the initial flux density. The radio blobs rapidly fade beyond the brightening zone. Figure 9 presents VLBI maps of SS433 obtained at 4.99 GHz using the European VLBI Network (EVN) at intervals of two days beginning on May 17, 1985 (Vermeulen et al. 1987). The dots mark two-day intervals in the trajectories of the propagating jets. On average, the pattern is very symmetrical, however deviations from symmetry in the two jets and deviations from the trajectory corresponding to the kinematic model are significant.

The radio emission along the VLBI jets is continuous, as is the jet activity in the optical, and generally obeys the regularities outlined above (weakening – brightening – weakening). However, the radio emission is also strongly modulated by the individual blobs, with characteristic times for their generation from one to several days. The modulation of the brightness of the VLBI jets seems appreciably stronger than the modulation of the Hα jets. However, in the

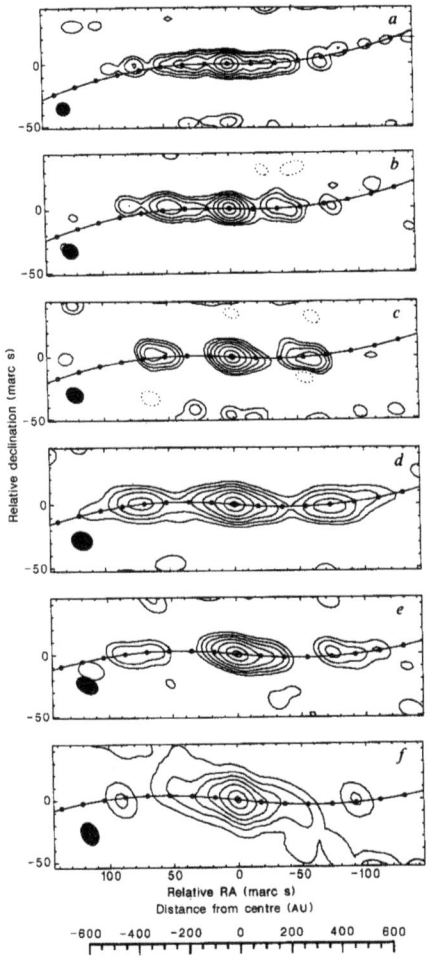

Figure 9. VLBI maps of SS433 (Vermeulen *et al.* 1987) obtained with an interval of 2 days beginning on May 17, 1985. The shaded ellipse corresponds to the angular resolution of the observations. The curves on each image show the trajectory of the jets according to the kinematic model, and the points mark two-day intervals on these trajectories for the propagating jets. The radio brightening zone is located approximately 50 mas from the centre. (Reproduced with permission from *Nature*.)

former case, we observe modulations in the images of the jets, while in the latter case, we have information about time variability of the intensity of the line emission. It is quite possible that the variability in the radio emission along the jets (in passive states of the object) is associated not with variability in the mass-loss rate in the jets, but with appreciable strengthening of the particle-generation process at certain phases of the nutational period; when the nutational and precessional motions add, for example. A similar effect for the optical jets has been well studied (see the previous section), but it is a projection effect, when the line emission adds up at a single radial velocity at phases of extrema in the nutation period, creating bright emission lines in the spectrum, given the name "bullets" in early studies. The optical line emission along the jets is essentially continuous, with the exception of modulations with a characteristic time scale of $0\overset{d}{.}3$–$0\overset{d}{.}5$. Accordingly, the mass-loss rate in the jets is likewise nearly continuous.

The "active" section of the VLBI jets begins about $1.5 \cdot 10^{14}$ cm or 2 mas from the center (Paragi et al. 1999) and ends just beyond the brightening zone at $\approx 4 \cdot 10^{15}$ cm. The active section of the Hα jets begins $1.7 \cdot 10^{14}$ cm from the center (the maximum emission arises at $4 \cdot 10^{14}$ cm; Borisov and Fabrika 1987) and ends just before the radio brightening zone. We can see that the same region in the jets gives rise to the most intense emission in both the radio and in the optical lines. It is likely that a single mechanism acts in this region (interaction of the jets with the slow wind), leading to the generation of synchrotron electrons and maintaining the clouds of relatively cool gas that radiate in the optical lines. This mechanism will be discussed in detail in the section "Structure and formation of the jets", and we describe it here only qualitatively.

The radio brightening zone is located at a distance of about $3.7 \cdot 10^{15}$ cm, which corresponds to 5.6 days of flight. Vermeulen et al. (1987) proposed that the interaction between the jets and the gas of the optical star's wind ceases at this distance from the center of SS433. The jets sweep out the gas of the slow wind on the surface of the precession cone, and the wind refills sections of the precessional cone after the passage of the jets. After the 164 days of the full precessional revolution, the jets pass through new gas. The jets move through new gas for only a few days, after which they enter a region that is free of wind, not having had time to be filled, where the jet gas clouds can expand freely and the radio emission can become stronger. This same scenario was discussed by Davidson and McCray (1980) as an explanation for the length of the optical jets. However, it follows from the observations that the optical jet emission ends at a distance of $3 \cdot 10^{15}$ cm, which corresponds to a radiation time of $\Delta t_j = 4.5$ days, one day less than the time of flight to the brightening zone. In addition, the optical star of SS433 overfills its critical Roche surface, and is unlikely to possess a powerful, isotropic wind. It has been proposed (Panferov and Fabrika 1997; Panferov 1999) that the jets interact with the wind from the accretion disk, and that the wind speed in near-polar regions of the disk in this scenario should be $V_w = (\Delta t_j/P_{pr})V_j \approx 2\,000$ km/s. Beyond the zone of interaction between the jets and wind, there is a zone in which the Hα clouds expand (the zone of expansion whose size comprises one day of flight of the jets). The expansion of the clouds leads to substantial variations in the jet structure and an increase in the efficiency with which the relativistic particles are generated. This mechanism is indeed able to explain the appearance of the radio brightening zone.

Equatorial Wind

The most recent radio interferometric observations of SS433 with high angular resolution (Paragi *et al.* 1999, 2000; Blundell *et al.* 2001) revealed unusual structure in the inner regions. Paragi *et al.* (1999) discovered a gap ~ 5 mas ($3.7 \cdot 10^{14}$ cm) in size in the radio emission at 1.6, 5 and 15 GHz. The size of the gap increases toward lower frequencies, as it roughly should in a conical jet geometry (Blandford and Königl 1979; Hjellming and Johnston 1988) if the weakening of the intensity in the central region is due to synchrotron self-absorption. The central source is located along the line connecting the jets, but its position is not symmetric relative to the two jets. The gap is larger on the side of the receding (western) jet. These geometrical properties, and also the different intensities and spectra of the radio emission of the two inner jets, clearly indicate that the inner radio jets are weakened by free–free absorption as well as synchrotron self-absorption. The size of the gap also varies with time, and the geometry of the absorbing gas probably depends on the precessional and orbital phases. The central gap in the radio emission testifies to the presence of an equatorial envelope. The absorbing gas surrounds the binary system in the form of an inclined disk-like envelope, whose projection onto the plane of the sky is roughly perpendicular to the direction of the jets.

Paragi et al. (1999) found clear evidence for the existence of gas in the plane perpendicular to the jets of SS433. Radio-emitting clouds were detected at 1.6 GHz at a distance of 40–50 mas (200–250 AU) on both sides of the central source, visible in Fig. 8c. Radio features in significant disagreement with the kinematic model for the jets ("anomalous ejections") were also observed in SS433 earlier (Romney et al. 1987; Spencer and Waggett 1984; Jowett and Spencer 1995). However, deviations of the jets by more than 5–10° have never been observed in the optical spectrum, in spite of the fact that the spectral observations cover an interval tens of times longer than VLBI observations. The equatorial regions of emission discovered by Paragi et al. (1999) have a very high brightness temperature (10^{7-8} K), which excludes the possibility that they are due to thermal radio emission. These regions have different shapes in different observing seasons (Paragi et al. 2000, 2002), i.e., they vary on characteristic time scales of at least tens of days. The available observations are insufficient to reveal any periodicity in the variations of these regions.

In the data of Blundell et al. (2001, 2002), the emission perpendicular to the jets does not form regions separated from the source, and instead forms a smooth halo-like structure. The radio spectra of these components are flat ($\alpha = -0.12 \pm 0.02$, $S_\nu \propto \nu^\alpha$), as is characteristic of thermal emission. However, the high brightness temperature is in complete contradiction with a thermal emission mechanism.

The speed of the equatorial wind in the immediate vicinity of SS433 has been derived from absorption-line radial velocities (Fabrika et al. 1997a). It depends on the angle above the plane of the precessing disk, and varies from ≈ 100 km/s to $\approx 1\,300$ km/s, with a mean speed of ≈ 350 km/s. The speed of expansion of this envelope has also been estimated from the proper motion of clumps in the VLBI images (Paragi et al. 2002) to be $\sim 1\,200$ km/s.

The presence of powerful gas streams in SS433 flowing outward in the plane of the binary system is confirmed by a whole series of independent observations in the optical, X-ray, and radio studies. We will return to an interpretation of the equatorial wind when describing other observations below. We will describe the structure of these flows in more detail when we discuss the supercritical accretion disk of SS433. The existence of such flows follows naturally from modern concepts about the formation of accretion disks in binary systems in which the donor star overfills its critical Roche surface.

W50

The well known radio nebula W50 surrounds SS433 on scales of tens of parsecs (Fig. 8a). A review of the results of studies of this nebula is given by Margon (1984). The central position of SS433, elongation of the nebula in the east–west direction (PA $\approx 100°$), along the axis of the jet precession cone, and many X-ray and optical data leave no doubt that W50 was formed (at least in this direction) as a result of interactions between the jets and the interstellar gas. The recent discovery of equatorial radio structure in SS433 (Paragi et al. 1999), which probably formed as a result of gas flowing from the system from the point L_2 beyond the accretion disk (Fabrika 1993), suggests that W50 could be excited by the constant activity of SS433 even in the direction perpendicular to the jets (north–south). A dense equatorial wind with velocity ≈ 300 km/s would be easily able to fill the body of W50 over $\sim 10^5$ yrs. Of course, this does not mean that we should exclude the possibility that the supernova explosion that gave birth to the relativistic star SS433 played a role in the formation of W50.

The image of W50 presented in Fig. 8a was obtained by Dubner et al. (1998) on the VLA from 1465 MHz continuum observations. It is reminiscent of a seashell. The central part of the nebula forms a nearly ideal circle with radius $29'$ (42 pc for a distance of 5.0 kpc). This could be a supernova remnant, however its size is not consistent with the standard surface brightness–diameter relation for remnants (Margon 1984). This may be associated with uncertainty in the distance to SS433.

Based on the observed HI-line morphology and an analysis of the interaction of W50 with the interstellar gas, Dubner et al. (1998)

found that the systemic radial velocity of the nebulosity is 42 km/s, which in turn leads to a kinematic distance of 3.0±0.2 kpc. However, the accepted distance of 5 kpc was derived by comparing the pattern of motions in the radio jets with the kinematic model, and there can be no doubt about the legitimacy of the latter. Thus, the distance to W50 derived both from the radial velocity of the nebula and from the surface brightness–diameter relation is appreciably smaller than the distance to SS433 derived from the known velocity of propagation of the optical jets, 0.26c, which, in turn, is measured using the transverse Doppler effect. This is a fairly difficult problem. We believe the solution lies in the unusual properties of W50.

The velocity of W50 obtained from the HI lines is in reasonable agreement with the systemic velocity of SS433 derived from the radial-velocity curve for the He II $\lambda 4686$ line (27±13 km/s, Crampton and Hutchings 1981a; -13 ± 12 km/s, Fabrika and Bychkova 1990). The modest discrepancy between the velocities of the nebula and the object could easily be explained as a result of the kick given to the object and the effect of uncompensated momentum during the supernova explosion in the binary system. The eastern and western optical filaments of W50 have radial velocities of 79 and 54 km/s, respectively, and the formal mean velocity is 67 ± 6 km/s (Mazeh et al. 1983). However, these data are likewise in disagreement with the long-baseline radio observations (and with the same kinematic model), as well as with recent CHANDRA X-ray observations of jet emission lines detected at a distance of $\sim 10^{17}$ cm from SS433 (Migliari et al. 2002), which demonstrate that the eastern jet is approaching the observer while the western jet is receding.

The plane of the Galaxy passes nearly perpendicular to the axis of W50 from its western side, making the western part of the nebula shorter and brighter (Fig. 8a). The two sides of the nebula are asymmetric in many respects; for example, the radio spectral index of the central region of W50 is $\alpha \approx 0.5$, while the eastern and western parts of the nebula have spectral indices of 0.8 and 0.4, respectively (Dubner et al. 1998). Dubner et al. (1998) estimated the total kinetic energy of expansion of the nebula to be $\sim 2 \cdot 10^{51}$ erg (for a distance of 3 kpc), which corresponds to a total flux of kinetic energy of $3 \cdot 10^{39}$ erg/s for a life time of $2 \cdot 10^4$ yrs for the nebula (Zealey et al. 1980).

The Extended Jets

The eastern and western optical filaments of the W50 nebula (Zealey et al. 1980; Kirshner and Chevalier 1980; Königl 1983; Mazeh et al. 1983) lie inside the projection of the jet precession cone at a distance of $R_{W50} \approx 50$ pc from SS433. The filaments are oriented perpendicular to the jet. Extended X-ray jets that can be traced out to the optical filaments have also been detected, with the maximum X-ray emission "fringing" these filaments (Watson et al. 1983). The optical spectra of the filaments show that the gas is heated by shock waves travelling at a speed of 50–90 km/s. The line intensity ratio S [II] $\lambda 6717/\lambda 6731$ has been used to estimate the electron density, $n_e \approx 10^2$ cm^{-3}, and gas pressure $P \approx 3 \cdot 10^{-10}$ erg cm^{-3} in the filaments (Königl 1983). Since the filaments are formed by the sweeping up of the interstellar gas by the jets, the pressure in the filaments should correspond to the dynamical pressure of the jets. Based on this idea, Königl (1983) and Fabrika and Borisov (1987) estimated the mass-loss rate in the SS433 jets, which, allowing for the real opening angle of the jets $\theta_j \approx 1°$, is $\dot{M}_j \sim 5 \cdot 10^{-7} M_\odot/$yr (corresponding to a kinetic luminosity $L_k \sim 10^{39}$ erg/s).

Knots of infrared emission were detected by the IRAS satellite at the western wing of W50 (Band 1987), located along the propagation axis of the western jet. The eastern wing does not have any appreciable infrared emission. The spectra of the knots in the four IRAS bands (from 12μ to 100μ) are fairly steep. In her mapping of W50 with the ISOCAM camera of the ISO observatory for studies of the processes decelerating the jets and their interactions, Fuchs (2002) detected a number of knots of emission in the $14-16\mu$ band, some of which coincide with regions of millimetre emission in the CO (1–0) transition at 115 GHz. It is possible that the western SS433 jet collides with and heats dusty regions, or it may be that the infrared emission is synchrotron radiation.

The large-scale jets of SS433 observed in W50 represent a unique laboratory for studies of the processes decelerating the jets and their interaction with the interstellar gas. The diffuse X-ray emission of SS433 (the X-ray lobes) was first detected by Seward et al. (1980). Watson et al. (1983) mapped the X-ray emission in the vicinity of SS433 using data obtained with the Einstein Observatory. The extended X-ray lobes or jets extend to the east and west from the source along the precession axis of the radio jets, in full agreement with the orientation and asymmetry of W50. The X-ray emission becomes appreciable about 20 pc (15′) from the source, reaches a maximum at a distance of 50 pc (in the region of the optical filaments) and disappears at distances of 60 pc. The diffuse X-ray emission is much softer than the emission from the central source. The luminosity of each jet is $\sim 6 \cdot 10^{34}$ erg/s (0.5–4.5 keV). Assuming that the X-ray emission was thermal, Watson et al. (1983) estimated the total thermal energy of the X-ray gas to be $\sim 1.2 \cdot 10^{51}$ erg, in reasonable agreement with the data of Dubner et al. (1998).

One striking feature of the extended X-ray jets is that their total opening angle is about 20° (Brinkmann et al. 1996), much smaller than the opening angle for the precession cone of the optical and inner radio jets (40°). The same geometry is shown by the outer radio lobes of W50 (Fig. 8a), whose total opening angle is appreciably smaller than that of the precession cone in the kinematic model. It would be natural to expect the kinematic model to be in agreement with the geometry of the extended structures in W50. In addition, the surface brightness of the X-ray emission grows with approach toward the precession axis, although, at first glance, it seems that the X-ray jets should be hollow and have their maximum radiation along the generating line of the precessional cone ($\pm 20°$).

To illustrate the extended jets, Fig. 10 presents an X-ray image of SS433 constructed using ASCA GIS data (Kotani 1998). This is one of the deepest X-ray images of the vicinity of SS433. This image was obtained by joining several images obtained with different time exposures. The central bright source is SS433, and the total size of the entire jet system is about one degree.

Subsequent investigations of the X-ray emission in the vicinity of SS433 were carried out using the ROSAT, ASCA and RXTE observatories (Yamauchi et al. 1994; Brinkmann et al. 1996; Safi-Harb and Oegelman 1997; Safi-Harb and Petre 1999). The X-ray spectra of the two extended jets are different. The spectrum of the eastern jet is non-thermal, with a power-law photon index $\Gamma \approx 1.6$. There is a region of thermal emission at the end of this jet ($T \approx 0.4$ keV), with the X-ray and radio structures being very similar. The emission of the western jet is appreciably softer ($\Gamma \geq 2.3$), and may even be

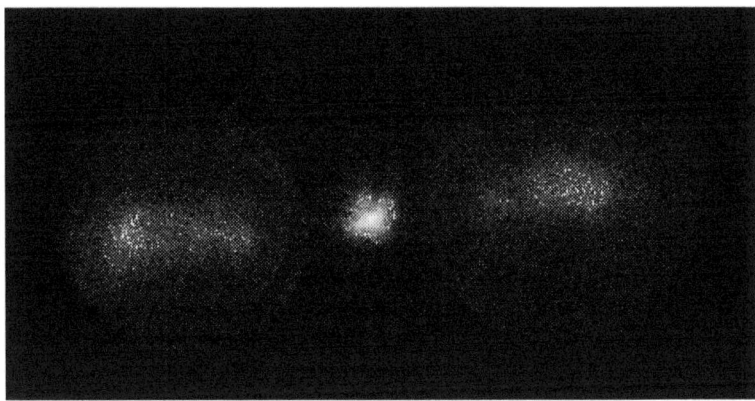

Figure 10. X-ray image of SS433 derived from ASCA GIS data (Kotani 1998). The central bright source is SS433, and the total size of the entire jet system is about one degree.

thermal, but no additional thermal emission is detected at the end of this jet (at the radio "ear"). The ROSAT data (Brinkmann et al. 1996) indicate that the X-ray structure does not vary significantly along the extended jets, although the contribution of the soft radiation surrounding the jet grows outside. Namiki et al. (2000) found that, with distance from the source, the X-ray spectrum becomes softer, and the spectrum is non-thermal and continuous, without emission lines. Analysis of RXTE observations (Safi-Harb and Petre 1999) confirmed that, in a broader energy range (to 100 keV), the spectrum of the extended jets (the eastern jet) is non-thermal, with $\Gamma \approx 1.45$, with the total X-ray luminosity of the jet being $\sim 1.2 \cdot 10^{35}$ erg/s. No γ radiation from W50 or SS433 has been detected (Geldzahler et al. 1989; Rowell 2001).

The increase in the brightness of the X-ray jets with approach toward the precession axis, as well as the decrease in the opening angle of the jet-propagation cone at large distances – the focusing of the extended jets – could be associated with hydrodynamical collimation of the precessing jets (Peter and Eichler 1993) or with the interaction of the jets with material from the supernova remnant and the formation of secondary reflected shock waves propagating inside the precession cone (Velazquez and Raga 2000). In this latter case, it is even possible to explain the helical structure observed in the radio ears of W50 (Dubner *et al.* 1998). The velocity of the shocks in the direction of the symmetry axis in this case should be no lower than ~ 130 km/s, so that a perturbation created at the precession cone surface reaches the axis in 10^5 yrs at a distance from the object of 40 pc.

However, it is not ruled out that the jet precession angle can change with time. In the slaved disk (precession) model, there is a beautiful mechanism that makes possible time variations in the precession angle of the donor star. Matese and Whitmire (1983, 1984) showed that a misalignment between the stellar rotational axis and the orbital axis could be enhanced in the mass transfer through the inner Lagrange point (the supernova explosion serves as a suitable mechanism initially introducing this misalignment). It is well-known that the orbit of the close binary system will tend to be circularized due to tidal friction (and mass transfer), and the misalignment between the rotational and orbital axes decreases with time. However, during the flow of gas through the inner Lagrange point, the specific angular momenta of the lost and remaining gas are imbalanced, so that the misalignment in the axes can be maintained (or even increased) after the orbit approaches circularity.

The alignment of the stellar and orbital axes due to tidal effects occurs over $\sim 10^5$ yrs (Papaloizou and Pringle 1983), which is less than the time for the star to evolve to the stage of overfilling its critical Roche lobe. Consequently, for the mechanism of Matese and Whitmire to operate, either the relativistic object in SS433 must form after the star has filled its Roche lobe, or this star should already be fairly evolved before the formation of the relativistic object. This conclusion is rather important. Indirect evidence that the mechanism of Matese and Whitmire operates is provided by the fact that the time for alignment of the stellar rotational and orbital axes in a binary system such as SS433 is less than or roughly equal to the time for circularization of the orbit (Papaloizou and Pringle 1982), and the orbit in this system is nearly circular, $e < 0.05$ (Fabrika *et al.* 1990). If the formation of a circular orbit in SS433 as a result of the supernova explosion is not purely coincidental (for example, the compensation by chance of the orbital angular momentum by the asymmetrical explosion of the supernova, which is unlikely), a mechanism that prevents the rapid alignment of the axes in this system is required.

The X-ray Jets

Early Observations

Early X-ray studies of the central source in SS433 are described in the review of Margon (1984). Beginning in the mid-1980s, thanks to the discovery of X-ray lines of the jets and observations during eclipses of the accretion disk, it has become clear that the X-ray emission is thermal, and is radiated primarily in inner regions, immediately above the accretion disk, by cooling gas of the relativistic jets on scales of $\sim 10^{12}$ cm. The total luminosity of the X-ray emission is $L_x \sim 3 \cdot 10^{35}$–10^{36} erg/s, substantially lower than the bolometric luminosity of the accretion disk, $L_{bol} \sim 10^{40}$ erg/s. The X-ray emission is strongly variable, and its intensity and spectrum (like those of the optical emission) depend on the activity state (flares), orientation of the disk and jets (precessional phase), and effects of eclipses of the optical star and absorption in the surrounding gas (orbital phase).

The X-ray iron lines were first detected in an EXOSAT spectrum of SS433 (Watson et al. 1986; Stewart et al. 1987; Brinkmann et al. 1988). These observations showed a relatively broad line that moved through the spectrum. This movement was in good agreement with the kinematic model if the line is emitted by highly ionised iron (Fe XXV, 6.7 keV) in the blue (approaching) jet; the emission itself was obviously thermal. The corresponding line from the receding jet had not been detected, and it was thought that this could be the result of eclipsing of the receding jet by the accretion disk (in which case the X-ray jet must be comparatively short) or of substantial weaking of the intensity of the receding jet due to the effects of relativistic aberration. It was also concluded that the X-ray gas of the jets had a low temperature ($kT \sim 2$ keV). The low emissivity of the X-ray gas corresponding to this temperature led to the requirement for a very high kinetic luminosity of the jets, $L_k \sim 10^{40-41}$ erg/s.

In subsequent GINGA observations of SS433 (Kawai et al. 1989; Brinkmann et al. 1991), X-ray brightness decreases were reliably identified with eclipses of the accretion disk. Some of these eclipses were in very good agreement with optical eclipses, even according to simultaneous optical observations (Goranskii et al. 1997). The eclipse light curves varied substantially depending on the orientation of the disk (precession phase).

The X-ray line emission of the jets was not resolved in the GINGA observations, but the complex behaviour of the broad shifted iron line at $E \approx 7$ keV had already been noted. This was described as precessional motion of "narrow" iron emission against a background of broad line emission. During eclipses, the intensity in the entire line decreased in proportional to the total flux, indicating that most or all of the X-ray emission was generated in the jets. The temperature of the emitting gas derived from the GINGA data decreased sharply during eclipses, from $kT \sim 30$ keV to $kT \sim 12$ keV at the centre of the eclipse, from which it follows that the temperature of the jets falls off with increasing distance from the source. In subsequent analyses of the GINGA data (Yuan et al. 1995), a narrow moving line component formed in the approaching jet was distinguished. The intensity of this component was approximately constant in the rest frame of the jet. It was noted that the intensity of the remaining broad iron-line component (a weakly ionised iron line or blend of many lines) varies in proportion to variations of the total X-ray flux, in accordance with the precessional variations of the orientation of the accretion disk. When the disk is maximally turned toward the observer, the object becomes brighter.

Localisation of the X-ray Source

The X-ray emission outside the binary system is relatively weak, since observations of the deepest eclipses in SS433 (Kotani 1998) indicate that the fraction of external radiation is less than 30%. More exact estimates are either not available, or they become model dependent: What fraction of the X-ray emission is formed in the uneclipsed, cooling jets ($\sim 10^{13}$ cm)? What fraction is reflected or re-emitted in gas moving in the wind? Is there an additional source of X-ray emission further from the system, in the region of maximum emission of the optical and radio jets ($\sim 10^{15}$ cm) and the radio brightening zone?

Recent CHANDRA HETGS observations (Marshall et al. 2002) detected extended X-ray emission around the central source on scales from $1'' - 2''$ ($(0.7-1.5) \cdot 10^{17}$ cm) to $6''$. Unfortunately, information about the structure of the central source itself was lost due to the pileup effect during these observations. The extended X-ray source is elongated in the direction of the jet precession axis, with its intensity growing toward the centre. The total luminosity of this source is $L_{x,ext} = 0.6\% L_x \approx 2 \cdot 10^{33}$ erg/s; Marshall et al. (2002) did not detect any emission lines in its spectrum.

The CHANDRA, ASIS–S observations reported by Migliari et al. (2002) confirm that the X-ray jets are resolved on scales of several arcseconds. The direction of the jets is completely consistent with the direction of the radio jets, and the maxima of the X-ray emission in the eastern and western jets are observed at distances of $\gtrsim 2 \cdot 10^{17}$ cm, however the central source of SS433 was likewise affected by pileup effects in these observations. The 2–10 keV X-ray luminosity at these distances from the source is $L_{x,ext} \approx 3\text{--}4 \cdot 10^{33}$ erg/s, which is about $\sim 3\%$ of the observed average X-ray luminosity of SS433.

Migliari et al. (2002) report the detection of emission lines shifted in accordance with the kinematic model. A line at ≈ 7.3 keV was found in the spectrum of the eastern (approaching) jet, while a line at ≈ 6.4 keV was found in the spectrum of the western (receding) jet. Both emission lines may be the Fe XXV Kβ (7.06 keV) iron line, shifted because of the gas motion in the jets with the velocity $0.26c$. The relative intensities of these two lines are also in agreement with the idea that the lines are formed in the jets. The time for the jets to travel to the emitting region is ~ 200 days. The emitting region is quite extended, and covers no less than one whole precessional cycle of the jets (Migliari et al. 2002). The continuum spectra (0.8–10 keV) of these two regions can be fit by a bremsstrahlung spectrum with a temperature of ~ 5 keV, however they can be equally well described with a power law with a photon index of 2.1 ± 0.2.

The extended X-ray emission detected by Migliari et al. (2002) cannot be radiation of the supercritical accretion disk scattered in external gas, since the spectra of the western and eastern components should be the same in that case. These data directly indicate reheating of the jets at distances between $3 \cdot 10^{15}$ cm (the end of the optical jets and the radio brightening zone) and $\sim 10^{17}$ cm.

It is likely that this X-ray emission on scales of arcseconds does not bear any relation to the extended X-ray jets described above, since the emission of the extended jets (deceleration of the jets) becomes appreciable at distances from the centre that are hundreds of times larger ($\approx 15'$). The X-ray emission on arcsecond scales could be associated with interactions between the jets and the disk wind, i.e., with the radio-emitting regions (the VLBI and VLA jets). Future X-ray observations with arcsecond and subarcsecond angular resolutions will provide answers to this question.

ASCA Data. Lines and Spectrum of the Jets

The ASCA observatory carried out about 30 observations of SS433 at various phases of the orbital and precessional periods (Kotani *et al.* 1994, 1996, 1997ab; Kotani 1998). The X-ray eclipses have various depths depending on the orientation of the accretion disk, which blocks from half to two-thirds of the radiation. ASCA observations were able to resolve the jet emission, and for the first time separate Fe XXV Kα, β, Fe XXVI Kα (appreciably weaker than the line of the helium-like ion), and Ni XXVII Kα lines detected for the approaching and receding jets. Weaker Kα lines of Mg XII, Si XII, Si XIV, S XV, S XVI, and Ar XVII were detected only from the approaching jet; many unresolved lines at 1–1.5 keV were also detected, as well as a fluorescent stationary line of neutral or weakly ionised iron Fe I–X at 6.4 keV (EW(Fe)$\approx$ 50 eV), which probably forms due to reprocessing of the radiation by the gas surrounding the jets and making up the wind of the accretion disk. Thus, a fundamentally new possibility arose of using the line intensities as a diagnostic for the X-ray jets of SS433. Future observations of the X-ray eclipses of the jets by the optical star of SS433 with better spectral resolution (such as that provided by CHANDRA) will open rich opportunities for direct investigations of the inner jets and the region above the photosphere of the wind in which the jets appear.

The spectrum at hard energies of 5–9 keV where the iron lines are emitted is in good agreement (taking into account both absorption and the contribution of the main lines) with a power-law with a photon index $\Gamma \approx 0.69$. At softer energies (1–4 keV), where lines of less heavy elements are emitted, the power-law spectral index is $\Gamma \approx 1.09$. The power-law approximations to the spectrum in these two intervals are consistent with the same absorption $N_H \approx 6.8 \cdot 10^{21}$ cm^{-3} (Kotani et al. 1996). The luminosity of SS433 at 2–8 keV is $L_x \approx 6.3 \cdot 10^{35}$ erg/s, with the luminosity in the brightest line of the approaching jet being $L(\text{Fe XXV K}\alpha)^- \approx 2.3 \cdot 10^{34}$ erg/s.

In a simple model of adiabatically cooling jets, the ratio of the intensities of the Fe XXV Kα and Fe XXVI Kα lines yield a temperature for the base of the jets of $kT_0 \approx 22$ kev. The temperature where the X-ray jets end – at a distance from the source of $(2-3) \cdot 10^{13}$ cm, where the gas becomes thermally unstable – is $\sim 1 \div 0.1$ keV. Beginning with a distance corresponding to a jet gas temperature of 6–7 keV, there is absorption or obscuration of the receding jet.

Estimates of the kinetic luminosity and mass-loss rate in the jets lead to fairly high values. However, nearly all such estimates have assumed improbably large opening angles for the jet, $\theta_j = 5°$, while the jets of SS433 are actually substantially better collimated than this, $\theta_j \approx 1°$ (Borisov and Fabrika 1987; Marshall et al. 2002). In their more detailed jet model, Brinkmann and Kawai (2000) estimated the kinetic energy flux to be $L_k \sim 5.7 \cdot 10^{39}$ erg/s. They demonstrated on the basis of ASCA observations that the jet emission lines could be used to study finer effects and to investigate the jet structure in detail.

Observations by the ASCA group (Kotani et al. 1997ab; Kotani 1998) led to heavy-element abundances appreciably higher than their solar values. To explain the spectra, the metal abundance must be enhanced by a factor of 1.5–2, and the abundance of Ni (Ni XXVII$^\pm$ at 7.3 and 7.7 keV) must be increased by more than a factor of 20. This result could have far-reaching implications, for example, about the occurrence of thermonuclear reactions at the surface of a *neutron star* inside the supercritical accretion disk. However, the observations of the optical jets are consistent with the idea that they have a normal chemical composition. The cited nickel abundance could be obtained if some additional effects were not taken into account in the adopted approximation of the SS433 X-ray spectrum, since the signal/noise ratio falls sharply at energies > 7 keV (Kotani et al. 1997ab). A model with ballistic jets cooling due to expansion and radiation was used, but it is likely that additional sources of heating must be included in the description of the X-ray jets (Brinkmann et al. 1988), for example, reheating by collimated radiation of the supercritical disk or shock processes arising when the jet gas exits the nozzle (a channel in the wind). Similar additional factors could affect the intensities of metal lines derived in models. The CHANDRA observations (Marshall et al. 2002) are not consistent with very high metal abundances.

ASCA Data. Equatorial Wind

The observations of the ASCA group revealed important behaviour that can be most successfully interpreted in terms of a picture with absorption of the radiation of the receding jet by a factor of 2–3, with the magnitude of the absorption growing with distance from the source (Kotani et al. 1996). The ratio of the intensities of the Fe XXV $K\alpha^+$/Fe XXV $K\alpha^-$ lines from the two jets was found to $\approx 0.24 \pm 0.06$, appreciably lower than the value 0.66 expected for the given precessional phase. The lines in the approaching jet should be brighter than those in the receding jet as a consequence of relativistic boosting (see the following section for more detail), but the lines of the receding jet proved to be systematically weaker than expected even after allowing for this effect. Thus, it is necessary to invoke additional absorption of the light from the receding jet. However, the object screening this jet cannot be the accretion disk, since the lines that are most subject to such screening would be the hottest lines emitting in the jets closer to the source, which is not the case. It was concluded that the systematic relation $(I^+/I^-)_{soft} < (I^+/I^-)_{hard}$ was valid for lines forming at various temperatures (Kotani et al. 1996, 1997ab; Kotani 1998). The observed line intensities required a weakening of the radiation from the most distant regions of the receding jet by a factor of two to three. This means that the "accretion disk" itself has relatively small dimensions, and our view of the receding jet is unobstructed at distances of $\sim 10^{12}$ cm from the source. Further, regions of the receding jet radiating at distances of $\sim 10^{13}$ cm (for comparison, the size of the system is about $a \approx (4-5) \cdot 10^{12}$ cm) begin to experience appreciable absorption.

Kotani et al. (1996) proposed that the absorption occurs in gas lost by the system through the outer Lagrange point L2 ("sprinkling" the disk). Possible observational manifestations of an intense loss of gas in SS433 through the point L2 are discussed by Fabrika (1993). It is likely that precisely such flows deform the orbital light curve in the optical (Zwitter et al. 1991; Fabrika 1993), and that their effect is observed on larger scales in the regions of radio emission perpendicular to the jets detected in VLBI images (Paragi et al. 1999; Blundell et al. 2001). It is possible that these equatorial regions can be detected most effectively in Hα line emission surrounding SS433 in the form of an extended disk illuminated by the precessing accretion disk, in which case we should expect variable extended Hα emission at distances of $\sim 1''$ from SS433. However, to our knowledge, such observations have not been carried out on the HST.

The results of ASCA X-ray observations are presented in the thesis of Kotani (1998). At the precessional phase when the disk is maximally facing the observer (near T_3 moment), the source is bright, the Fe K edge is deep, the receding jet undergoes appreciable absorption, and the temperature of the gas is higher in the approaching jet than in the receding jet. At precessional phases when the disk is viewed edge-on (near $T_{1,2}$), the X-ray source is weak, the Fe K edge is shallow, and the intensities of the emission lines of both jets are the same. This last circumstance is very important, and makes the conclusions that can be drawn firm.

CHANDRA Data. Narrow Multi-temperature Jets

CHANDRA observations of SS433 (Marshall et al. 2002) have mainly confirmed the conclusions reached earlier on the basis of the ASCA observations. The CHANDRA observations are especially important for our understanding of the X-ray jets of SS433, since the excellent spectral resolution enables the direct detection and verification of the effects described above. The spectrum of SS433 obtained with the CHANDRA HETG spectrometer (Marshall et al. 2002) is shown in Fig. 11, where the range of wavelengths in Angstroms corresponds to 1.08–8.3 keV. It was possible to detect more than 20 emission lines from the approaching jet, 6 lines from the receding jet and a line of neutral (or weakly ionised) iron at 6.42 keV in the spectrum. The strongest line was a helium-like line of iron Fe XXV. The multi-temperature model for the cooling jets was confirmed; lines of the lighter elements Ne and Mg ($T \sim 10^7$ K) are observed together with hot lines of Fe and Ni ($T \sim 10^8$ K). An appreciable number of lines at low energies were not identified. The overlapping of these lines could make a significant contribution to the continuum.

The fact that CHANDRA observations were able to resolve the jet lines in SS433 is a very important result. These lines proved to be appreciably broadened, FWHM $\approx 1\,700$ km/s, with the line widths and radial velocities having approximately the same magnitude independent of the temperature of the radiation. Marshall et al. (2002) found that the opening angle for the X-ray jets is $\theta_{j,x} = 1°\!.23 \pm 0°\!.06$. Recall that the opening angle of the optical jets found by Borisov and Fabrika (1987) was $\theta_{j,opt} = 1°\!.0$–$1°\!.4$. When modeling the profiles of the moving Hα lines (Borisov and Fabrika 1987), the intensity distribution across the jet was represented as a two-dimensional Gaussian with $\sigma = \theta_j/2$. In contrast to the "short" X-ray jets, for which fairly simple geometrical considerations suffice for estimates of the opening angle, estimation of the opening angle of the optical jets requires modeling of the line profiles, since nutational and precessional shifts contribute to the total line widths. The coincidence of the opening angles for the X-ray and optical jets is remarkable. This means that the SS433 jets are indeed conical and move along strictly ballistical trajectories, from the source itself at distances of $\sim 10^{11}$ cm (where the jets emerge from below the wind photosphere and the temperature of the jet gas is $\sim 10^8$ K) to the beginning of the zone of expansion of the Hα clouds, $\approx 3 \cdot 10^{15}$ cm (where the temperature of the jet gas is $\approx (1-2) \cdot 10^4$ K).

Based on the positions of the lines, Marshall et al. (2002) deduced that the velocity of the X-ray jets was $\beta = 0.2699 \pm 0.0007$, which is 2920 ± 440 km/s higher than the velocity of the jets in the kinematic model derived from optical data (Margon and Anderson 1989). However, if we make the comparison using refined parameters of the kinematic model (Eikenberry et al. 2001), the formal

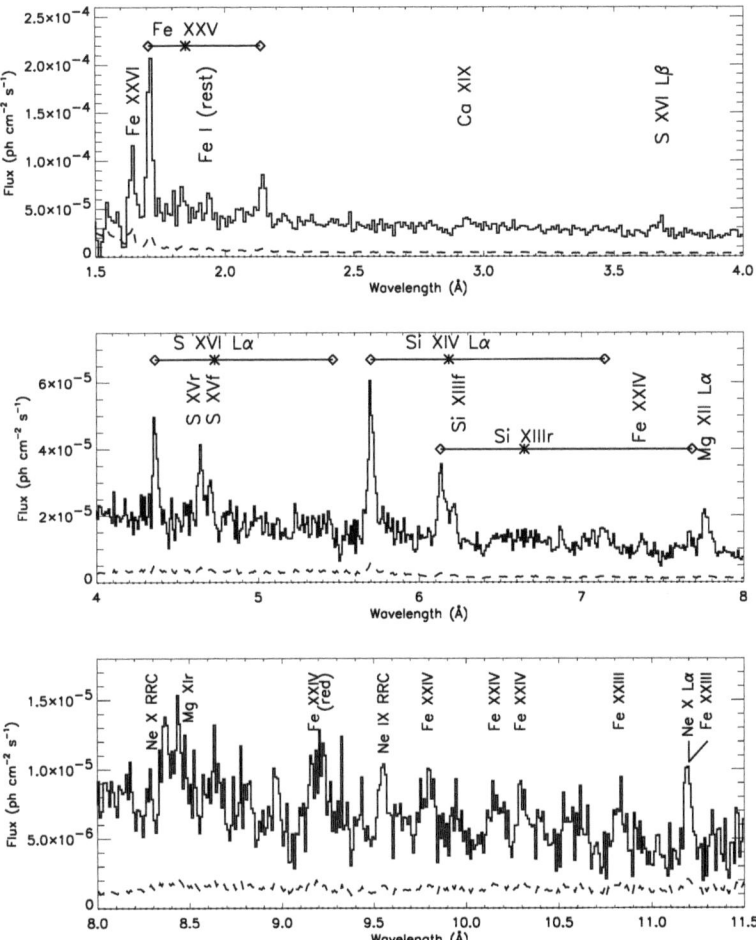

Figure 11. CHANDRA HETG spectrum of SS433 (Marshall *et al.* 2002). We can see primarily lines corresponding to the blue jet, which are marked. The horizontal bars join the same lines radiating in the blue and red jets (diamonds), while the asterisks mark the unshifted positions of these lines. The dotted line shows the statistical uncertainty.

difference in the velocities is reduced to 1560 ± 340 km/s. It is important to note here that the temporal instability of the jet velocity (jitter) can reach $\pm3\,000$–$5\,000$ km/s. Given the relatively short duration of the CHANDRA observations of SS433 (29 ks), it seems premature to conclude that the X-ray and optical jets propagate with different speeds. The complete coincidence of the collimation angles for the X-ray and optical jets, in turn, suggests that they correspond to the same physical object observed at different stages of its evolution.

The 0.8–8 keV CHANDRA X-ray spectrum is in good agreement with a power law ($\Gamma = 1.35$, $N_H = 9.5 \cdot 10^{21}$ cm^{-3}) and with the spectrum obtained by ASCA (Kotani et al. 1996); the mean luminosity of SS433 at 2–10 keV is $L_x = 3.1 \cdot 10^{35}$ erg/s. The gas temperature can be determined fairly reliably based on the ratios of the line fluxes of hydrogen-like and helium-like ions. The observed jet emission lines form at temperatures from $1 \cdot 10^8$ K to $5 \cdot 10^6$ K. The density-sensitive Si XIII triplet can be used to derive the electron density, $\sim 10^{14}$ cm^{-3}, of the gas in the region of the jet where the temperature is $1.3 \cdot 10^7$ K. Very weak lines (Ne X, Ne IX) arising as a consequence of radiative recombination have been detected. If the strength of these lines is due to photoionisation, this requires a luminosity of order $L_x \sim 10^{40}$ erg/s, which is, in principle, plausible if the X-ray radiation of SS433 is collimated along the jets. However, based on the absence of other strong lines that should arise during photoionisation, Marshall et al. (2002) concluded that the gas was collisionally heated.

The emission measures for various ions have been determined in a model with a conical, adiabatically cooling jet; optically thin thermal and collisional plasma; and normal elemental abundances. A four-component (four-temperature) model was also constructed, in which the temperature of the jet gas falls from $1.1 \cdot 10^8$ K to $6 \cdot 10^6$ K and the electron density falls from $2 \cdot 10^{15}$ to $4 \cdot 10^{13}$ cm^{-3} at distances from the base of the jet from $2 \cdot 10^{10}$ to $2 \cdot 10^{11}$ cm. The kinetic luminosity of the jets derived in this model is $L_k \sim 3 \cdot 10^{38}$ erg/s. Thus, the X-ray jet inferred by Marshall et al. (2002) proves to be very short. This places substantial constraints on the distance from the relativistic star at which the bases of the jets are located, such that there are no instantaneous effects of eclipsing of the jets by the optical star.

The X-ray iron line at 6.4 keV is not resolved in the CHANDRA spectra of Marshall *et al.* (2002), FWHM < 1 000 km/s. This line most likely arises via fluorescence in the cool wind gas (Kotani 1998), or even in a cocoon around the base of the jets (Fabrika 1997). Based on an analysis of the X-ray eclipses and eclipses of the He II $\lambda 4686$ emission line and the times of egress of the X-ray source and He II source from behind the limb of the optical star, Goranskii *et al.* (1997) concluded that the region of He II emission surrounds the X-ray source in the SS433 accretion disk. We will consider the structure of the disk separately.

The CHANDRA data confirm the screening of the receding jet detected by ASCA – the radiation of this jet is substantially weaker than that of the approaching jet. However, the temperatures of the two jets proved to be approximately equal. More prolonged observations are needed to test the hypothesis that the radiation of the receding jet is absorbed in material flowing from the system, since the gas flowing outward in the plane of the disk could be appreciably non-uniform in the azimuthal direction.

Marshall *et al.* (2002) noted an interesting coincidence between the expansion velocity of the jets in the transverse direction (more precisely, the maximum possible expansion velocity) derived from the line widths and the sound speed at a temperature of $\sim 10^8$ K derived from the line intensities. If the width of the jets is determined by the free expansion of the gas at the base of the jets, the opening angle should be equal to $\theta_{j,x} = 2c_s/V_j$, where c_s is the sound speed for protons and V_j is the jet velocity. The jet opening angle that is obtained for a gas temperature at the base of the jet of $T_0 = 1.1 \cdot 10^8$ K is $\theta_{j,x} \approx 1°.4$, which virtually coincides with the value derived from the observations. As the distance from the centre increases, the gas cools, the sound speed falls, and the jets become strictly ballistic. This coincidence of the opening angles represents a weighty argument in support of the idea that the temperature T_0 is measured just at the point where the jets emerge from under the photosphere of the cocoon surrounding their base. On its own, this does not shed light on the jet collimation mechanism, which is most likely hydrodynamic. However, we can conclude that the jets should initially (in inner regions hidden from the observer) be collimated no more poorly than is observed in the X-ray and optical, and that the operation of the collimation mechanism should end somewhere just before the emergence of the jets from under the photosphere.

In recent observations with the CHANDRA HETGS published by Namiki et al. (2003) SS433 was observed in precessional phase "disk edge-on". The authors have found that a width of the iron line Fe XXV Kα (FWHM(Fe) $\sim$ 4900 km/s) is considerably greater than that of silicon line Si XIII Kα (FWHM(Si) $\sim$ 2000 km/s). Marshall et al. (2002) have also noted such a trend, that the width of lower energy lines are slightly less than average width of all lines studied in the spectrum, however this trend was only marginal in their data. In the spectra of Namiki et al. (2003) the silicon line width is in agreement with the line widths found by Marshall et al. (2002), but the iron line is notably broader. The authors suggested that they detected a progressive jet collimation along its axis. These new data show that it is necessary to accumulate more observations with high spectral resolution (CHANDRA) at different precessional phases and also during the accretion disk eclipses to understand the structure of the SS433 X-ray jets and their distance of the source. Probably, a Compton scattering in the jet gas or in surrounding medium may play an important role in broadenning of the X-ray spectral lines. Furthermore at the disk orientation "edge-on" its inner parts (the jet bases) are obscured partly by outer rim of the disk (the section "The Supercritical Accretion Disk and the Components from the Photometric Data"), that is the geometrical effects have to influence the X-ray spectrum.

In very recent observations with the gamma-ray observatory INTEGRAL Cherepashchuk et al. (2003) detected a hard X-ray radiation of SS433 in the energy band 20–100 keV. The hard X-ray spectrum appears relatively flat in this band with the power-law photon index $\Gamma \sim 2$. The luminosity of SS433 in the hard X-rays is $L_x \sim 3 \cdot 10^{35}$ erg/s (25–50 keV) and $L_x \sim 1.2 \cdot 10^{35}$ erg/s (50–100 keV). Cherepashchuk et al. (2003) found a precessional variability of the hard X-ray flux: when the disk is maximally facing the observer, the flux in the 25–50 keV band increases more than two times comparing with the precessional orientation "edge-on". The precessional variability confirms the notion that the outer rim of the disk blocks the inner region at the disk orientation "edge-on". Note that the relativistic boosting effect changes the flux depending on the precession phase with an amplitude about ≈ 40% if we consider only one jet, and the effect is quite weak (9%) for the both antiparallel jets.

The presence of the hard power-law component in the X-ray spectrum means in turn a Comptonisation of soft X-ray photons generated in the inner disk (in X-ray jets) on relativistic electrons. The relativistic particles may be accelerated in the same inner jets, where the jets leave out the funnel of the supercritical disk (the next section).

Inhomogeneity of the Jets and X-ray Variability

As a rule, investigators analysing X-ray observations of the jets have assumed that the jets are conical and completely filled with gas. In contrast, the filling factor of the optical jets must be quite small (Davidson and McCray 1980; Begelman *et al.* 1980); the filling factor for the jet clouds at the distance of the maximum Hα emission ($\approx 4 \cdot 10^{14}$ cm) is $\sim 10^{-6}$ (Panferov and Fabrika 1997). The jet gas is collected in clouds by the action of thermal instabilities. It is quite likely that the region of formation of these clumps (Bodo *et al.* 1988; Brinkmann *et al.* 1988; Kotani *et al.* 1996) is located at the end of the X-ray jets, where the gas cools to a temperature of ~ 0.1 keV and should begin to fragment (see following section). The cloud sizes predicted by the thermal-instability mechanism (Brinkmann *et al.* 1988) and derived from relative hydrogen line intensities (Panferov and Fabrika 1997) are $\sim 10^8$ cm. There should be thousands of such clouds even in the relatively short X-ray jets, and it is very unlikely that it will be possible to detect their presence in the X-ray, for example, from X-ray variability.

The structure of the Hα line profiles in the optical jets suggests the presence of $\sim 10^3$ larger-scale formations that could be considered clusters of clouds (Borisov and Fabrika 1987; Panferov and Fabrika 1997), with the characteristic time for the formation of these structures being $\sim 10^2$ s. This approximately corresponds to the time for the motion of the gas along the X-ray section of the jets. Thus, the structure of the optical jets and the relatively modest size of the X-ray jets suggest that the X-ray flux is very likely variable on time scales corresponding to the formation times of these structures.

The time of ~ 100 s is also close to the time for the propagation of the SS433 jet beneath the photosphere of the wind of the supercritical accretion disk. For a mass-loss rate in the SS433 wind of ~ 10^{-4} $M_\odot$/yr, the wind-photosphere radius is R_{ph} ~ 10^{12} cm (van den Heuvel 1981; Lipunov and Shakura 1982; Fabrika 1997). This size is in good agreement with the dimensions of the source of optical and UV radiation around the relativistic star derived from observations, $R_{UV,opt} = (1.5-2) \cdot 10^{12}$ cm (Dolan et al. 1997). If we identify this object with the opaque part of the wind or with a channel in the wind, the time for gas moving with the velocity of the jets to cross this region is 200–300 s. The jets should be accelerated and collimated on this, or even shorter, time scales (if the geometry of the inner region is complex and the photosphere where the X-ray emission emerges is smaller than the photosphere where the UV radiation is generated). We can also conclude based on the coincidence of the time for the generation of large-scale inhomogeneities in the jet and the time for the propagation of the jet beneath the photosphere that we should expect X-ray variability with a characteristic time scale of hundreds of seconds.

Kotani et al. (2002), Safi-Harb and Kotani (2002) recently detected such variability in PCA/RXTE observations obtained during the active state of SS433. Strong stochastic variations of the brightness of SS433 are observed at energies of 2–10 and 10–20 keV on time scales of $10^2 - 10^3$ s, with the minimum variability time scale being 50 s. Although the accretion disk was viewed edge-on during these observations, i.e., its orientation was not amenable to variability studies, this variability was reliably detected. It is likely that the X-ray flux of SS433 is also variable in quiescent periods, possibly with a lower amplitude than in periods of activity.

Chakrabarti et al. (2002) discussed plausible mechanisms for the production of bullet-like ejecta with a time scale of 50–100 s in the accretion disk of SS433 as a possible explanation for the observed short-time-scale variability. They suggest non-linear oscillations of shocks in the accretion disk or close to a sound barrier in the sub-Keplerian accretion flow as one such mechanism. In this case, the accretion rate will exhibit appreciable modulations with the time-scale needed.

Variability of SS433 on short time scales (< 100 s) has been searched for in the UV at 1400–3000 ÅÅ in HSP/HST observations (Dolan et al. 1997). An upper limit to the variability amplitude of < 1.2% was derived from these observations. We can hope that short-timescale variability of SS433 will eventually be detected in the blue and UV, where the emission is radiated in regions of the wind close to the jets or by gas surrounding the place where the jets emerge. Like in the X-ray, the question of detecting rapid variability in the UV probably reduces to the sensitivity of the detectors used and the signal/noise ratio of the observations. The presence of optical variability on time scales of several minutes is well established (for example, Goranskii et al. 1987; Zwitter et al. 1991); this is stochastic variability with an amplitude of $\sim 0^{m}\!.1$ that does not disappear even during eclipses of the accretion disk.

Structure and Formation of the Jets

In the previous sections, we described the main observational data on the SS433 jets, as well as constraints on the physical conditions in the jets that can be derived directly from these observations. Here, we describe the physical conditions and state of the gas in the jets in more detail (some of these results will naturally be model-dependent), as well as the main published theories concerning mechanisms for the jet formation. Our review is not aimed at a complete treatment of various theoretical questions.

The State of the Gas in the Optical Jets

The precessional motion of the jets presents excellent opportunities for studies of the physical state of the gas in the jets. Comparing the intensities of the hydrogen emission lines at various angles of the two jets to the line of sight and assuming that the two jets are intrinsically identical and that their internal properties do not depend on orientation (precessional phase), we can characterise the gas heating mechanism and elucidate the structure of the radiating clouds of gas. Differing absorption in the stellar wind of the emission from the "+" and "−" optical jets (in contrast to the X-ray jets) is ruled out, since the size of the jets is two orders of magnitude larger than the dimensions of the SS433 binary system.

The observed intensity of the jet spectral lines $\mathcal{J}_{obs}$ will depend on the relativistic aberration of the emitted radiation. In a jet with a finite length or consisting of individual evolving fragments with finite life times (Lind and Blandford 1985; Begelman et al. 1984; Panferov and Fabrika 1997), the intensity of the line radiation in the rest frame of the jet is $\mathcal{J}_{co} = (1+z)^3 \mathcal{J}_{obs}$, where $z = \delta\lambda/\lambda_0$ is negative for the approaching jet and, as before, $\lambda = \lambda_0 \gamma(1 - \beta\cos\eta)$. If we consider the continuum radiation, the exponent for this dependence should be replaced by $2 + \alpha$, where α is the spectral index ($I_\nu \propto \nu^{-\alpha}$). At time T_3 ($\psi = 0$, the jet axis is inclined at the minimum angle to the line of sight, 57°), the line emission from the approaching jet should be brighter than that from the receding jet, $\mathcal{J}_{obs}^- / \mathcal{J}_{obs}^+ \approx 2.4$. At precessional phases $T_{1,2}$ (when the jets are located in the plane of the sky), the intensities of the two jets should be the same.

Asadullaev and Cherepashchuk (1986) discovered that the jet radiation is anisotropic. They found that the ratio of the intensities of the two jets in the Hα line in a frame co-moving with the jets is maximum at precessional phase $\psi \approx 0$ and is equal to about 2–3. In this case, the anisotropy can be interpreted in a model with clouds in the jets, with the maximum radiation occurring in their frontal regions.

In their study of the behaviour of the intensities of the moving Hα lines based on a substantially larger amount of observational data, Panferov et al. (1997) found that the angular distribution of the jet radiation is rather complex. Figure 12 presents the precessional dependence of the intensities of the main components of the moving Hα$^\pm$ lines in a frame co-moving with the jets in units of 10^{-10} erg/cm² s. The vertical axis plots the angle between the "−" jet and the line of sight (the error in the values of this angle appearing in this same figure presented by Panferov et al. (1997) has been corrected). The intensities have been corrected for various values of the interstellar absorption ($A_V = 7\rlap{.}^{\mathrm{m}}3 - 8\rlap{.}^{\mathrm{m}}3$) near the true value for SS433, $A_V \approx 8$. The data have been obtained for the main components of the moving-line profiles (the "young" jet), which form near the source. This same figure (Figs. 12e, f) shows the behaviour

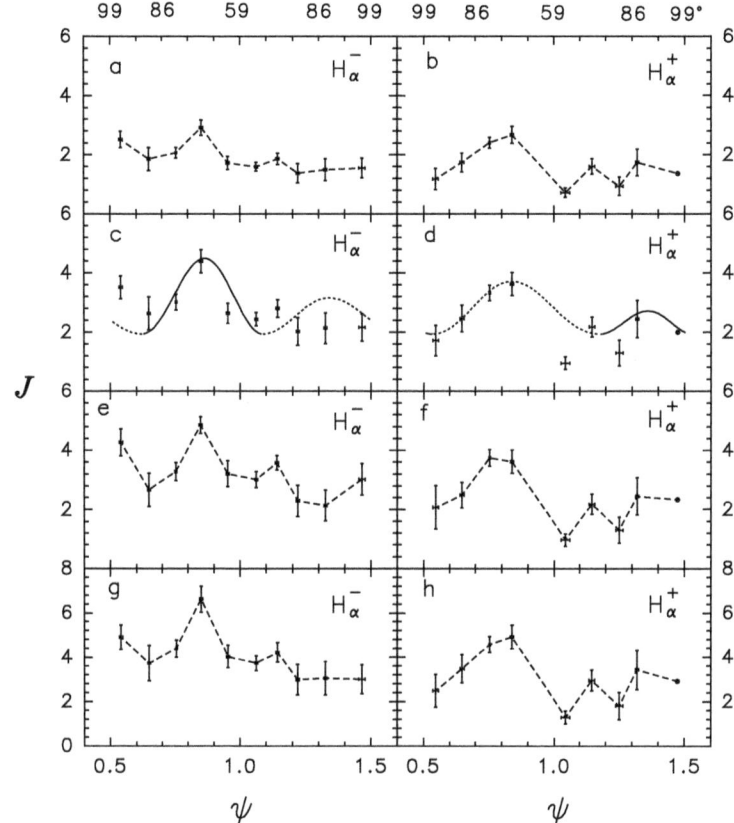

Figure 12. Precessional dependences of the intensities of the main components of the moving H$\alpha^\pm$ lines in a comoving coordinate frame in units of 10^{-10} erg/cm^2 s (Panferov et al. 1997). The upper axis gives the angle between the "−" jet and the line of sight. The intensities have been corrected for interstellar absorption using the values $A_V = 7^{\rm m}\!.3$ in (a) and (b), $7^{\rm m}\!.8$ in (c)–(f), and $8^{\rm m}\!.3$ in (g) and (h). Panels (e) and (f) show the intensities obtained by summing all the components of the moving-line profiles. Model intensity curves are shown in panels (c) and (d), where the solid curve corresponds to the radiation from the front hemisphere of the emitter and the dashed curve to radiation from the rear hemisphere.

of the intensities obtained by summing all components of the moving-line profiles. The figure presents the errors in the mean intensities in intervals of the precessional phase $\Delta\psi = 0.1$. Each phase interval contains the results of several authors obtained during various precessional cycles during the ten years in which observations have been made. Possible variations in the line intensities associated with active periods are not reflected in the data in Fig. 12, since the mean data were obtained over a long time interval. Possible inaccuracies in the adopted magnitudes of the absorption and of the projection effects described above (the nutational motion gives rise to an increase in the intensity of the moving lines at extrema of the radial-velocity curves), likewise, do not appreciably change the form of the dependences. In particular, the Hα light curves of the jets derived from the main component or from the entire profile are the same.

Figure 12 implies that the jet line radiation is anisotropic, and that the directional beams for the radiation of both jets are similar; there are maxima in the radiation both in the direction of motion and in the opposite direction (forward and backward), with the axis of the beam not coinciding with the jet-velocity vector. The maximum radiation is observed not at phase $\psi = 0$, but shifted to phases 0.8–0.9. Model intensity curves obtained taking relativistic aberration into account are shown in Figs. 12c and d, where the solid and dashed curves correspond to the radiation of the front and back hemispheres of the model radiator (a gas cloud in the jet was specified as a flattened spheroid). Panferov et al. (1997) found that the front sides of the clouds in the SS433 jets are a factor of ≈ 1.7 brighter than the back sides. The apparent equality of the intensities of the "+" and "−" jets is the result of relativistic aberration of the emitted light. The directions of the maximum radiation of the two jets deviate significantly from the direction of the jet motion, by $\sim 30-40°$, towards the direction of the precessional motion. The dynamical interaction of the jets with the surrounding medium could lead to such effects. Consequently, the dissipation of the kinetic energy of the jets is the dominant process in heating the jets.

Anisotropic radiation by individual clouds is possible if their optical depth in the hydrogen lines exceeds unity. Either these clouds are flattened or they are transparent; i.e., the probability for line radiation to exit is enhanced in a particular direction. This direction could correspond to the direction of shocks in the jets that make the gas transparent to line radiation due to the development of velocity gradients in the gas clouds. The jets of SS433 do not propagate through a previously established channel as they move through the surrounding medium, as would jets propagating in a constant direction, but must continually rebuild this channel, pushing it toward the direction of the precessional motion. The jets interacting with the wind create a cocoon and carry with them nearby pieces of the wind. The transverse density and velocity profiles of the inter-cloud gas in the jets will not be axially symmetric: larger gradients will develop in the direction of the precessional rotation, and it is from that direction that gas can flow into the jet. In this case, the direction of propagation of shock waves in the jet is inclined to the velocity vector toward the direction of the precessional rotation. The radiation of the jet gas in optically thick lines is maximum along this axis. The vector of the maximum Hα radiation of the clouds is closest to the line of sight at precessional phases $\psi \approx 0.8$. Either the inter-cloud gas flows relative to the clouds inside the jet or weak shocks move through the inter-cloud gas and clouds, but these perturbations propagate at an angle to the jet axis and arise on the side in which the precessional motion is directed.

Panferov and Fabrika (1997) studied the Balmer decrements in the SS433 jets. The relative intensities of the H$\alpha^{\pm}$, H$\beta^{\pm}$ and H$\gamma^{\pm}$ lines were found using a uniform dataset obtained over ten years of observations, deriving the individual decrements from spectra taken during a single night. The ratios of the hydrogen-line intensities are the same in the two jets, but vary appreciably with the precessional phase. At the phases of the maximum intensities of the moving lines ($\psi = 0.7 - 0.9$), Hα/H$\beta = 1.6 \pm 0.2$ and Hγ/H$\beta = 0.7 \pm 0.1$, while Hα/H$\beta = 0.9 \pm 0.2$ and Hγ/H$\beta = 0.8 \pm 0.1$ at the phases of the minimum intensities ($\psi = 0.0 - 0.2$). Such decrements are characteristic of high-density gas, $n_e \gtrsim 10^{12}$ cm^{-3}, when the populations of atomic levels are determined primarily by collisional processes. In addition, the effects of optical depth and projection are important during the formation of the jet hydrogen lines. A comparison of the observed relative intensities with the computational results of Drake and Ulrich (1980), which were obtained for a uniform layer of gas within broad intervals of the physical parameters took into account the influence of the Stark effect on the probability for the photons to exit their emission region, demonstrates that the gas in the SS433 jets should be dense and be formed of compact clouds. The effective size of individual gas clouds, or, more precisely, the optical depth in the Hα line, varies appreciably depending on the orientation of the jets. The precessional phases when bright H$\alpha^\pm$ lines are observed correspond to small optical depths of the layer. The following parameters of the gas clouds and jets have been derived:

- mean particle density $n \approx 10^{13}$ cm^{-3},
- gas temperature $T_e \approx 2 \cdot 10^4$ K,
- optical depth of the layer to line radiation $\tau(\text{H}\alpha) \sim 10^2$–$10^4$, depending on the jet orientation,
- size of the clouds $l \sim 10^8$ cm,
- number of clouds in the jet $\sim 10^{12}$,
- volume filling factor for the jet clouds $\xi \sim 4 \cdot 10^{-6}$,
- kinetic luminosity of the optical jets $L_k \approx 10^{39}$ erg/s.

Estimates of the kinetic luminosity for two intervals of the precessional phase (bright and weak jet lines) are in reasonable agreement. These were derived from the ratio $\epsilon_{H\beta}/n$ – the ratio of the mean efficiency of the radiation of a unit volume of gas in the Hβ line to the gas density (Drake and Ulrich 1980). The kinetic energy flux is $L_k \approx L_{H\beta} \, m_p \, n \, V_j^3 / 2 \, \epsilon_{H\beta} \, R_j$, where the jet luminosity in this line is (on average) L$_{H\beta} \approx 7 \cdot 10^{35}$ erg/s, the mean quantity $\epsilon_{H\beta} \approx 1$ erg/cm^3 s, and $R_j \approx 4 \cdot 10^{14}$ cm is the length of the jets corresponding to the maximum line radiation.

The Zones of Sweeping out and Expansion

Clouds of the supposed dimensions freely expand on a characteristic time scale of $l/c_s \sim 100$ s, much shorter than the lifetimes of the moving-line components (4 days). Consequently, the clouds are not in a freely expanding state in the jets, and there must be some mechanism preventing their expansion. The jets sweep up the wind gas along the surface of the precessional cone; the length of the swept-out zone is $P_{pr} V_w \sim 3 \cdot 10^{15}$ cm if the disk wind speed in circumpolar regions of the disk ($\pm 2\,\theta_{pr} = 40°$) is $V_w \approx 1\,500 - 2\,000$ km/s. Within this region, the Hα clouds are protected from disruption by the dynamical pressure of the surrounding gas, and they can slowly evolve. The Hα/Hβ intensity ratio is appreciably higher for the weaker secondary moving-line components that form in distant sections of the jets than for the main components (Panferov and Fabrika 1997). This testifies to a decrease in the density of the clouds in the jet with time (with increasing distance from the source). If the wind density depends on distance as $\hat{n} \propto R^{-2}$, the density of the clouds that are prevented from expanding by the dynamical pressure of this gas falls by nearly two orders of magnitude by the end of the swept-out zone. The dimensions of the clouds are $l \propto n^{-1/3} \propto \hat{n}^{-1/3} \propto R^{2/3}$, and therefore increase approximately fivefold. The volume emission measure in the gas of the clouds $n^2 l^3$ along the optical jets falls as R^{-2}, and the volume filling factor of the jets falls as $\xi \propto R^{-1}$.

Beyond the zone in which the wind gas is swept up, the jets move in gas-free space, since the gas there was swept up during previous precessional passages of the jets, and the slow wind has not had time to refill this space with gas. Here, the dynamical pressure of the gas on the clouds is much lower, and the clouds are reheated and expand – the radiation in the hydrogen lines ceases.

The zone of expansion of the clouds begins at the boundary of the sweeping-out zone, at a distance of about $3 \cdot 10^{15}$ cm (4.5 days of flight of the gas). Here, the density of the incoming gas decreases sharply. The clouds, which are no longer confined beyond the sweeping-out zone, expand and fill the entire volume of the jets. This process probably leads to the appearance of numerous shocks in the jets and to the rapid heating and expansion of the gas clouds, and the efficiency of the generation of relativistic particles is increased. Recalculating the cloud parameters to the outer boundary of the sweeping-out zone, Panferov and Fabrika (1997) found that the freely expanding clouds can fill the entire jet volume in approximately one day. This is consistent with the location of the well known brightening zone in the radio jets of SS433 (see the section "The Radio Jets and W50"),

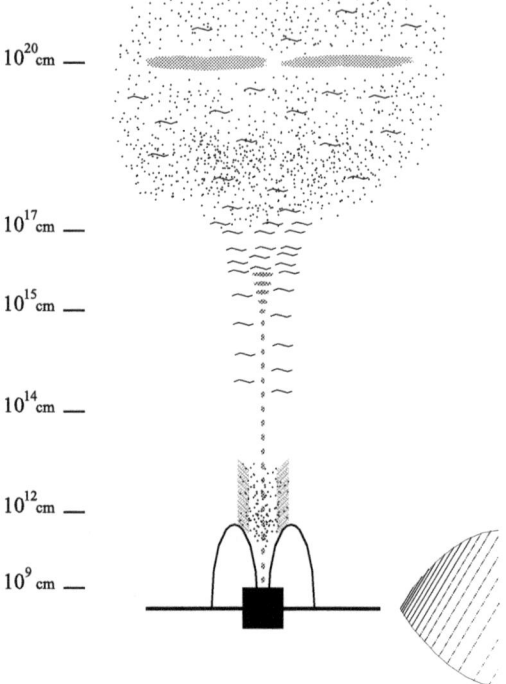

Figure 13. Schematic of the SS433 binary system and jets. The regions of X-ray emission of the jets are shown by points, and the regions of radio emission by wavy lines. The solid curves mark the photosphere of the slow wind. The cocoons of hot gas above the wind photosphere are shaded. The optical filaments at the ends of the large-scale jets ($\sim 10^{20}$ cm) are also shown. Note that the jets on scales of $\sim 10^{12-13}$ cm are the main source of the X-ray radiation. The black square shows a region $\sim 100 R_g$ in size in the supercritical accretion disk. The outer edge of the accretion disk and gaseous flows are not shown in this diagram.

which is observed in VLBI images (Romney et al. 1987; Vermeulen et al. 1987, 1993b) at a distance of $3.7 \cdot 10^{15}$ cm (5.6 days of flight of the jet gas), i.e., at a distance of $0.7 \cdot 10^{15}$ cm from the outer boundary of the sweeping-out zone.

Figure 13 presents a schematic of the binary system and the jets of SS433. The vertical scale indicated in the figure is logarithmic, but the scale itself is not continuous, since the purpose of the figure is to illustrate the main components of the jets described above. Regions of X-ray emission of the jets are shown by dots, and regions of radio emission by wavy lines. The cocoons of hot gas above the wind photosphere are shaded (see "The Supercritical Accretion Disk from Spectroscopic Data"). The optical filaments at the ends of the large-scale jets ($\sim 10^{20}$ cm) are also depicted. The jets are not observed below the wind photosphere, but they are the main source of X-ray emission on scales $\sim 10^{12-13}$ cm. The filled square shows the region $\sim 100 R_g$ in size in the supercritical accretion disk that has been studied using hydrodynamical simulations (see below). The schematic does not show either the outer edge of the accretion disk or gas flows, which in reality have an appreciable impact on the brightness of the system.

Heating of the Jets

The gas in the optical jet must be continuously reheated, since the radiative cooling time for the gas in the clouds is several orders of magnitude shorter than the time of their motion. The heating of the Hα clouds by UV and X-ray emission arising during the dissipation of internal shocks (Fabian and Rees 1979) requires dispersion velocities in the shocks of $\Delta V \sim 100$ km/s (Begelman *et al.* 1980) in order to provide agreement with the observed relative intensities of the helium lines in the spectra. However, this heating mechanism is improbable, as, in addition, it should lead to the appearance of an appreciable X-ray flux (that is not eclipsed, since the size of the jets is two to three orders of magnitude larger than the size of the binary system). In particular, if we recalculate the estimated X-ray luminosity obtained by Begelman *et al.* (1980) taking into account the now known distance, jet opening angle, and jet luminosity in the Hα line, $L_{H\alpha} \approx 1 \cdot 10^{36}$ erg/s in a frame that is comoving with the jets (Panferov and Fabrika 1997), the predicted jet luminosity at 2–10 keV is $\sim 10^{36}$ erg/s, which is approximately an order of magnitude higher than is acceptable.

In principle, heating of the jets by the collimated radiation coming from the accretion disk funnel is possible (Bodo et al. 1985; Fabrika and Borisov 1987; Panferov and Fabrika 1993), but this is always treated as being hypothetical, since we cannot see the direct radiation from the funnel due to the orientation of the SS433 system. Heating of the Hα clouds by UV radiation is unlikely, since the clouds should be uniformly heated, and the side of the clouds nearest the source would not be brighter than the forward side. Therefore, in the case of UV heating, the radial density of the clouds must be $N_H < 10^{18}$ cm^{-2}, while analysis of the hydrogen-line intensities implies hydrogen densities for the Hα clouds $N_H \sim 10^{21}$ cm^{-2}. For this same reason, in the case of heating by collimated X-ray radiation, the luminosity of the funnel at energies $\epsilon > 1.5$ keV must be no lower than $L_{x,c} \gtrsim 3 \cdot 10^{39}$ erg/s, in order to provide the required Hα flux (Panferov and Fabrika 1993, 1997).

The possibility of heating of the jets by collimated radiation from the funnel raises intriguing possibilities for investigating the directional beam of the funnel radiation. The precessional and nutational motions of the axis of the collimated radiation result in the "slow" jet (compared to the speed of light) exiting from the light-cone directional beam at some distance from the source. This will diminish the intensity of the heating, which could be detected via analysis of the profiles of the moving lines. In particular, the model profiles of the Hα$^-$ line shown in Fig. 7 were obtained precisely under the assumption that the gas in the jet is heated by collimated radiation (Panferov and Fabrika 1993). The beam of the radiation was represented both as a two-dimensional Gaussian and as a flat function (the radiation intensity does not depend on direction within the cone). It was concluded that, in the case of heating by collimated radiation, the total opening angle of the collimation cone was no less than $\theta_c > 14°$. However, the profiles of the moving lines proved to be insensitive to the form of the directional beam for the collimated radiation. It is known that the Hα$^\pm$ intensity falls off exponentially along the jet (Borisov and Fabrika 1987), and that this is probably determined by the gradual variation in the size of the clouds and in the emissivity of the gas along the jet. Moreover, there are direct data indicating that the dominant source of heating of the gas making up the clouds is interactions of the jets with the gas of the slow wind. Therefore, if the clouds are heated by collimated radiation as well, the contribution of this heating to the overall heat balance of the gas is not the dominant one.

Heating of the Hα clouds via interactions of the jets with the surrounding gas (Davidson and McCray 1980; Begelman et al. 1980; Brown et al. 1991; Panferov and Fabrika 1997) is currently the most promising mechanism. The parameters of the gas clouds in the jets will also be determined by interactions of the jets and wind. This mechanism is in good consistency with the observational data – the dependence of the intensity of the jet line radiation on the precessional phase and the anisotropy of the cloud radiation in the hydrogen lines. This mechanism is based on the necessary condition that the jets propagate through the accretion-disk wind, and makes it possible to understand the length of the optical jets and the formation of the radio brightening zone.

In their article based on the earliest spectroscopic data for SS433, Davidson and McCray (1980) derived surprisingly precise values for the main parameters of the jets and gas clouds: the electron density ($n_e \sim 10^{13}$ cm), temperature ($T_e \sim 1.5 \cdot 10^4$ K), size ($l \sim 10^8$ cm), and volume filling factor for the jet clouds. The requirement that the jet volume not be completely filled with radiating gas follows from energetic considerations. In order for the mass of the jets (and the kinetic-luminosity flux) to not be unreasonably high, the emissivity of the radiating gas (which is $\propto n_e^2$) must be rather high, and the clouds must not be too opaque to the line radiation ($\tau \propto ln$). The highest line emissivity of the gas is reached when there are a large number of relatively small and dense clouds. Davidson and McCray (1980), as well as Bodo et al. (1985), proposed that cool clouds are dissipated due to their interaction with the hot ($\sim 10^8$ K) intercloud gas that forms in the jets during their interaction with the wind. Brown et al. (1991) considered various mechanisms for heating the gas in detail, and concluded that heating via collisional interaction between the jets and wind was most likely. In this case, the line radiation of the gas should be linearly polarized (Brown and Fletcher 1992). The direction of the polarization is orthogonal to the jets, and the expected degree of polarization is $\approx 0.2\%$.

It is improbable that each individual cloud radiating in the Hα line experiences dynamical pressure from the gas incoming at the speed of propagation of the jet. Interactions at a relative velocity of $0.26\,c$ could easily lead to overheating and dissipation of the clouds. It appears that a more complex interaction is realised, when the first set of clouds to pass through a region sweep out the wind gas, possibly becoming disrupted in the process. The jets create a channel, which moves in the wind gas in accordance with the nutational and precessional motions. The surrounding gas could become entrained in the jets and flow into them. It is possible that shock waves or density waves propagate along the jets in the reverse direction with relatively low speeds, heating the clouds and preventing them from expanding. The parameters of the optical clouds, probably, as well as those of the fragments into which the X-ray jets are divided (Brinkmann et al. 1988), will be determined by complex processes occurring over the course of the evolution of the jets.

Based on long series of spectral observations, Kopylov et al. (1986) derived a limit on the deceleration of the jets of $\Delta V_j/V_j \lesssim 10^{-2}$ over 3–4 days in the lifetimes of the blobs. The loss of jet kinetic energy corresponding to this deceleration is

$$\hat{L}_k = 2 L_k \Delta V_j/V_j \lesssim 2 \cdot 10^{37} \, L_{k39} \text{ erg/s},$$

where L_{k39} is the kinetic luminosity in units of 10^{39} erg/s. Given the Hα luminosity of the jets $\sim 10^{36}$ erg/s, we find that the efficiency of cooling the gas via Hα radiation is $\gtrsim 0.05$, which is quite acceptable (Begelman et al. 1980; Panferov and Fabrika 1993).

Based on the condition that momentum be conserved,

$$\Delta \dot{M}/\dot{M} = \Delta V_j/V_j,$$

in order to satisfy the limit on the jet deceleration, the density of the decelerating gas must be

$$\hat{n} = 8 \Delta V_j L_k / \pi m_p \sin^2 \theta_j R_j^2 V_j^4 \lesssim 1.5 \cdot 10^5 \, L_{k39} \text{ cm}^{-3}.$$

It is obvious that the source of this gas must be the wind flowing from the accretion disk of SS433. The disk loses gas at a rate $\dot{M}_w \sim 10^{-4} \, M_\odot/\text{yr}$, and the wind speed in circumpolar regions is $V_w \gtrsim 1\,500$ km/s (Fabrika 1997). If this flow were isotropic, the density of the gas through which the jet passes would be

$$\hat{n} = \dot{M}_w/4\pi R_j^2 V_w m_p \approx 4 \cdot 10^6 \,\text{cm}^{-3}.$$

Naturally, this outflow from the SS433 disk should be anisotropic, with the wind in circumpolar regions having higher speeds and lower densities.

Ejection of Gas in the Jets

In the first years after the discovery of SS433, a number of authors (see, for example, Calvani and Nobili 1981) considered a model in which the gas was accelerated by radiation pressure and the flow was collimated into jets within the funnel of a thick accretion disk. Thermal instabilities in the cooling gas of the jets can naturally explain the formation of clumps of gas (Davidson and McCray 1980; Bodo *et al.* 1985), which are further observed as Hα clouds in the optical jets. Alternative approaches to the problem of the non-uniformity of the optical jets are possible. For example, Brown *et al.* (1995) proposed the existence of "radiative instabilities" in the jets, when the cool gas radiating in the Hα line does not form at all in isolated large-scale sections of the jets due to variations in the balance of heating and cooling processes. Now, in the light of X-ray studies of the jets, it is clear that the hot base of the jets is as non-uniform as the optical jets, and the need for such models is removed. However, this suggestion of Brown *et al.* (1995) remains very pertinent, and some "SS433-like" objects may not show prominent jet activity at all. Precisely the presence of a funnel in the supercritical accretion disk, in which the gas is not only accelerated, but also compressed into a thin stream, creates the conditions required for subsequent fragmentation of the jets. The presence of a disk wind with which the jets subsequently begin to interact due to their precessional and nutational motions creates unique conditions enabling these fragments to survive over the entire extent of the optical jets.

Bodo *et al.* (1985) considered the acceleration of the gas in the funnel of a thick accretion disk (Jaroszynski *et al.* 1980; Rees *et al.* 1982; Ferrari *et al.* 1985; Icke 1989). The funnel and disk are separated by the walls of the funnel, which is a dynamical structure.

Under the action of a number of effects, the wall material can enter the funnel and be efficiently accelerated outward by radiation pressure. The structure of the accretion-disk funnel has been considered in a number of studies (Lynden-Bell 1978; Sikora 1981; Narayan *et al.* 1983), as a rule, in the context of accretion disks around supermassive black holes. The funnel in a thick disk can naturally give rise to collimated radiation. Taking into account the effects of reflection of the radiation off the walls (Madau 1988) enhances the degree of collimation of the radiation exiting along the funnel axis. If we allow for the scattering of the wall radiation on the rarefied gas in the funnel ($\sigma_T \sim 1$), the degree of collimation of the radiation could prove to be rather high, especially if this gas moves outward along the funnel axis with the velocity $v_j \sim 10^{10}$ cm/s.

Via hydrodynamical calculations of supercritical disk accretion onto a black hole in a binary system assuming conditions close to those for SS433, Eggum *et al.* (1985, 1988) concluded that the funnel with its dynamical walls is formed in the inner regions, at several tens of gravitational radii. The opening angle of the funnel turns out to be $\theta_c \sim 30°$. The optically thin gas in the funnel is accelerated by radiation pressure to velocities $\sim 10^{10}$ cm/s. The walls consist of accreting gas, and bound the region of the funnel (photocone of the collimated radiation) from the region of convective plasma motions. Up to 80% of the released gravitational energy is accreted onto the black hole in the form of a kinetic energy flux and radiation. About 1% of the accreting material is ejected in the funnel (in the jets), yielding a value that is close to the relative mass-loss rate in SS433 jets, where $\dot{M}_j/\dot{M}_0 \sim 5 \cdot 10^{-7}/10^{-4} \sim 0.5\%$ of the mass reprocessed by the disk is lost in the jets. Computations have been done for accretion rates $\dot{M}/\dot{M}_{crit} = 1-10$ (where $\dot{M}_{crit}$ is the critical accretion rate required to produce the Eddington luminosity), while the accretion rate in SS433 reaches $\sim 10^3$. A more precise comparison of computational results with the observed picture for SS433 requires appropriate specification of the rate at which mass enters the outer edge of the disk, the mass-loss rate on scales of the spherisation radius, and, accordingly, the accretion rate in inner parts of the disk, which are difficult to determine on the basis of observational data. Overall, the computations of Eggum *et al.* (1985, 1988) illustrate a basic scheme according to which we can understand the mechanism for the ejection of gas and formation of jets in SS433.

The two-dimensional hydrodynamical computations of Okuda (2002, and references therein) support the mechanism of Eggum et al. (1985, 1988) for the formation of the funnel and dynamical walls in inner regions from several tens to hundreds of gravitational radii, where the radiation pressure exceeds the gas pressure. The rarefied gas is accelerated in the funnel to velocities of 0.1–0.2 c. The structure of the funnel can be very complex. The presence of convectively unstable zones located in a broad torus-like region on the other side of the dynamical walls of the funnel is also confirmed. Close to the black hole, there are advective plasma motions. Okuda (2002) proposed that the acceleration of the gas in the funnel could be stabilised by some mechanism which, under the conditions leading to the given funnel structure, results in a constant outflow speed of 0.26 c.

Comparatively recently, the significant role of convection has been recognized in gas dynamics and energy transfer in the inner regions of accretion disks (Stone et al. 1999; Abramowicz et al. 2002 and references therein). The convective motions can form very powerful gas outflows from the accretion disks, while, at the same time, the accretion flow is advective very close to the black hole. A strong wind from the accretion disk could also been formed because of standing shocks arising near a centrifugal barrier in the disk (Molteni et al. 1994; Chattopadhyay and Chakrabarti 2002). The simulations of Molteni et al. (1994) show that the disk wind propagates not only in circumpolar regions of the disk, but also in quite far directions from the disk axis ($\theta \sim 50° - 60°$). In particular, the very same wind structure is observed in SS433 (Fabrika 1997).

Hydrodynamical computations of supercritical disk accretion are very promising for improving our understanding of the mechanisms acting in SS433. It is likely that, in the near future, computations extending to scales $\sim 10^9 - 10^{10}$ cm, encompassing the spherisation radius of the disk, will be carried out, enabling the elucidation of how the supercritical disk wind forms. The spherisation radius of the accretion disk is (Shakura and Sunyaev 1973) $R_{sp} = GM_x \dot{M}_0 / L_e$, where $\dot{M}_0$ is the rate at which gas enters the disk, and M_x and L_e are the mass of the compact star and its corresponding critical luminosity. If the accretion rate at the outer edge of the disk is $\sim 10^{-4} M_\odot$ /yr, the spherisation radius is $R_{sp} \sim 10^{10}$ cm. Computations of the funnel and wind on larger scales, $\sim 10^{11} - 10^{12}$ cm, may make it possible to elucidate how the jets are collimated.

A number of models have been constructed for the formation of the jets and radiation of SS433 both in the funnel of a thick accretion disk and in a channel in the wind or gaseous envelope (Fukue 1987ab; Inoue et al. 2001), where the channel has shapes varying from conical to those in which the cross section of the channel grows with distance more slowly, $S \propto r$. Models with a channel or funnel in the accretion disk (or in material flowing outward from the disk) and collimated radiation in this channel are very appropriate for SS433.

It is interesting that models with a supercritical accretion disk around a neutron star in which the star does not possess a strong magnetic field (Okuda and Fujita 2000) give approximately the same results, since the funnel (as in the case of a black hole) forms around the rotational axis of the disk. An obvious difference is the accretion rate onto the surface of the neutron star and, accordingly, the rate at which gas is ejected from the inner regions of the disk, since, in contrast to a black hole, a neutron star cannot accept more material at its surface than the amount determined by the critical accretion rate.

Several mechanisms for the ejection of gas and formation of jets for the case of a neutron star with a strong magnetic field have been proposed. Lipunov and Shakura (1982) considered supercritical accretion onto a slowly rotating neutron star, where gas in the magnetosphere falls onto the surface along force lines and is ejected in clumps along the magnetic poles of the star with characteristic time intervals equal to the free-fall time for blobs of gas arriving from the magnetosphere. In the cauldron model of Begelman and Rees (1984), the jets are accelerated when they pass through the magnetopause, as though through narrow de Laval nozzles stabilised by the pressure of the neutron star's magnetic field. As in the previous model, an appreciable source of energy is required for the initial acceleration of the gas in the magnetosphere, which could plausibly be provided by the non-stationary accretion of blobs of gas or the rapid rotation of a young neutron star with an oblique magnetic field.

Arav and Begelman (1992, 1993) developed a model for the acceleration and collimation of gas in SS433 in which the jets are ejected relatively close to the neutron-star surface and propagate in a channel formed in the dense atmosphere created by the accreting gas. The boundary layer of this radiation-dominated channel in the atmosphere and the evolution of the channel and jet with distance from the source were considered in detail. The boundary layer is represented by a cocoon of low-density gas around the jets. It was demonstrated that the channel and cocoon exhibit collimating properties and efficiently collimate the radiation from the central regions. This radiation-dominated channel model could prove to be very useful for our understanding of the properties of channels in supercritical accretion disks and jet flows within them. In spite of the number of excellent models for SS433 as an object containing a neutron star, the presence of this one in SS433 is very improbable (see the next section).

Acceleration, Collimation and Fragmentation

The most promising models are those with the acceleration and collimation of gas in the funnel of a supercritical accretion disk around a black hole. Bodo *et al.* (1985) initially assumed that the radiation in the funnel was collimated, the gas was optically thin and the main mechanism scattering the radiation was Thomson scattering. Hot gas diffuses inside the funnel and is accelerated by the existing collimated radiation. The accelerated material acquires its essentially final speed while still deep inside the channel, at distances $r \lesssim 10^9$ cm, and a supersonic gas flow with a temperature $T \gtrsim 10^7$ K exits the funnel. Depending on the total luminosity of the radiation coming out of the funnel and the funnel opening angle, the final speed of the material is $\beta = 0.1 - 0.6$. In particular, for approximate values of the luminosity in units of the Eddington luminosity and the funnel opening angle appropriate for SS433, $L_c/L_e = 10$ and $\theta_c = 30°$ and $40°$, outflow speeds of $\beta = 0.35$ and 0.25 are reached.

If the following mechanisms:

(i) finer tuning of the velocity of motion to the value $0.26\,c$,
(ii) collimation of the flow along the propagation axis,
(iii) the formation of dense clumps of gas along the axis ($\theta_j \sim 0.02$),

act in an already accelerated flow of gas (for example, within the funnel in its upper part or in the region where the material exits the funnel), then, generally speaking, we will obtain jets like those in SS433. We consider each of these points in more detail below.

(i) Models for the acceleration of the gas in the SS433 jets by radiation pressure propose the action of "line-locking" (Milgrom 1979b; Pekarevich et al. 1984; Shapiro et al. 1986; Katz 1987). This mechanism was actively discussed at the beginning of the 1970s in connection with the acceleration of gas in envelopes surrounding active galactic nuclei and the formation of absorption lines in quasars. In the case of SS433, the observed jet velocity of $0.26\,c$ corresponds to the Doppler shift of the frequency of the main $L\alpha$ transition of the hydrogen atom to the frequency of the ionisation threshold, Lc. If acceleration by radiation pressure occurs via the absorption of photons in $L\alpha$ frequencies and there is an absorption edge Lc in the source spectrum, gas can be efficiently accelerated only to velocities $\approx 0.26\,c$, since the source radiation intensity falls off sharply beyond the Lc edge. This mechanism operates equally effectively in locking the $K\alpha$–Kc radiation of hydrogen-like and helium-like ions.

Investigations of the line-locking mechanism as applied to SS433 were carried out primarily before the discovery of the X-ray jets, and, accordingly, before it became clear that the main lines in these jets belong to hydrogen-like and helium-line ions of heavy elements. Therefore, consideration of heavy elements in these works had a somewhat speculative nature, which was fully justified in subsequent years. It was concluded that line-locking could be efficient in SS433, but with the additional proviso that either there was an enhanced content of heavy elements (such as iron) in the SS433 jets, or the jets were accelerated by collimated radiation. This second hypothesis now appears very natural.

The efficiency of line-locking under the conditions in SS433 has been placed under doubt more than once, but, as a rule, on the basis of substantially overestimated values for the kinetic luminosity of the jets. Nevertheless, it is difficult to imagine another acceleration mechanism providing the striking constancy of the jet velocity in SS433. The jet velocity does not depend on anything: the velocity remains constant, even in the presence of variations in the luminosity of SS433 during flares (by nearly an order of magnitude), in active states or in times of appreciable brightness decreases. The jets sometimes disappear (see the section "The Radio Jets and W50"), but they always reappear with the same velocity, $0.26\,c$. More generally, one can say that, when the SS433 jets contain cool gas clouds emitting in the hydrogen lines, the velocity of the jets is always the same.

The surprising constancy of the jet velocity provides a very strong argument in support of the line-locking mechanism. Even if the acceleration of the jets is not entirely due to this mechanism, it seems very likely that at least the tuning of the flow velocity to the observed value must be due to this effect. If the limiting velocity of $0.26\,c$ is approached not from below, via the acceleration of material in the funnel, but from above, via the deceleration of material, line-locking can stop the deceleration at the required velocity, since the gas begins to be accelerated by Lc photons when the velocity is decreased to the critical value $V_j = 0.26\,c$. If the flow of gas in the funnel accelerated in the inner regions to velocities $\sim 10^{10}$ cm/s begins to be decelerated by the wind coming in from the funnel walls (the gas in this wind can adhere to inhomogeneities in the main flow), this deceleration is sharply slowed at the threshold velocity $0.26\,c$ by the effect of line-locking. The mechanism for the formation of instabilities in material accelerated by radiation pressure proposed by Katz (1987) could sharply increase the efficiency of the formation of inhomogeneities in the accelerated flow.

(ii) One candidate for the collimation mechanism could be hydrodynamical collimation of the flow (Peter and Eichler 1996) due to interaction with the walls in the upper part of the funnel or with the walls of the gas cocoon surrounding the place where the jets emerge. Indicators of such a cocoon in SS433 would be the fluorescence of a "Fe I gas" made up of weakly ionised iron, manifest in the X-ray spectra (previous section), and evidence for a He II cocoon in the observed optical spectra ("The Supercritical Accretion Disk from Spectroscopic Data"). This same hydrodynamical collimation mechanism has been proposed to explain the narrowness of the extended X-ray jets associated with the W50 nebulosity ("The Radio Jets and W50").

The flow of gas in the funnel should interact with the walls, consisting of gas from the slow, dense wind. A wind may blow from the walls and collide with the main flow. In addition, due to the development of hydrodynamical instabilities such as Kelvin–Helmholtz instabilities, inhomogeneities (waves) should form in the walls and be carried away by the main flow. Interaction with the inhomogeneities and with the wind from the walls leads to the formation of axially symmetrical shocks in the flow, which move along the main flow of gas toward the funnel axis and are carried by the flow. The velocity of waves perpendicular to the motion of the main flow is equal to or slightly higher than the sound speed, $c_s \sim 10^8$ cm/s for a temperature $T \sim 10^8$ K (the temperature inferred for the base of the jets using ASCA and CHANDRA data). In spite of the fact that the flow itself is substantially supersonic, $V_j/c_s \sim 100$, a shock wave propagating along the flow toward the axis could be weak. Thus, a "soft" oblique shock could propagate along the flow, compressing gas along the axis of the motion. The collimation angle of this compressed region is $2c_s/V_j \sim 0.02$ (or somewhat higher, since the temperature inside the funnel should be higher than the temperature inferred from the observed base of the X-ray jets), i.e., the gas is compressed into a narrow jet. The question is whether thermal instabilities are capable of confining the gas in this jet via the sharp decrease in the pressure due to the radiative cooling, or whether the compressed gas on the funnel axis remains too hot to form dense clouds. Nevertheless, the gas on the channel axis should cool very efficiently due to its enhanced density, and its temperature should fall. In any case, if the temperature at the exit from the funnel is lowered to the measured value $T_{j,x} = 1 \cdot 10^8$ K, the jets cannot expand for kinematic reasons, and are "frozen" at the level $\theta_j = 1°\!.4$, as discussed in the previous section. Further, the thermal instabilities can divide the jets into dense clumps, but this will occur inside the cone θ_j and after the exit from the funnel, when the gas temperature is lowered to $\sim 10^6$ K by adiabatic expansion and radiative losses in the X-ray jets. As far as we are aware, no hydrodynamical computations of the motion of the flow in the upper part of the funnel have been carried out.

Lebedev *et al.* (2002) presented results of laboratory experiments, where hypersonic plasma jets are generated in conically convergent flows. The convergent flows are created by electrodynamic acceleration of plasma in conically divergent array of fine metallic wires (Z–pinch array). Stagnation of plasma flow on the axis of symmetry forms a standing conical shock effectively collimating the flow in

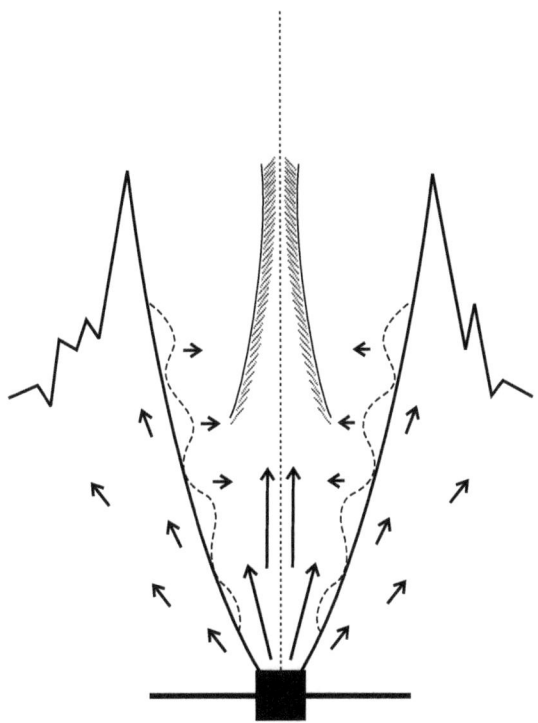

Figure 14. Schematic of the funnel in the disk wind. As in Fig. 13, the black square denotes the region in which hydrodynamical modelling of the supercritical disk has been carried out (Eggum et al. 1985, 1988; Okuda 2002). The bold curve denotes the photosphere of the slow wind. The dotted curve shows inhomogeneities arising in the walls of the channel. Proposed oblique shocks compressing gas into a thin ($2c_s/V_j \sim 0.02$) stream are shaded.

the axial direction. In the experiments a hypersonic (M $\gtrsim$ 20) well–collimated jet is generated. The jet collimation depends obviously on radiative cooling rate of the shocked plasma on the axis of symmetry. The experiments show high efficiency of the discussed collimation mechanism, the geometry of the device (Lebedev et al. 2002) is very similar to that of the funnel in SS433, which we discuss. The jets formation by the convergence of supersonic conical flows was considered and developed by Cantó et al. (1988) in application to formation of interstellar jets.

Figure 14 shows a schematic of the funnel in the disk wind illustrating the features discussed above. As in the previous figure, the filled square denotes the region in which hydrodynamical modeling of the supercritical accretion disk has been performed (Eggum et al. 1985, 1988; Okuda 2002). The bold curve denotes the photosphere of the slow wind. Inhomogeneities arising in the funnel walls can appear on the boundary of the interaction of the fast flow and slow wind. The proposed oblique shock waves compressing gas into a narrow ($2c_s/V_j \sim 0.02$) jet are shaded.

At the gas compression on the axis of jet in the conically convergent supersonic flows a generation of relativistic electrons is quite probable. In principle, the mechanism of acceleration of the relativistic particles may not differ on the particle acceleration mechanism in radio jets (the section "The Radio Jets and W50"), where the electrons are accelerated in shocks originating in interaction of the jets and the disk wind. The relativistic electron energy needed for formation of the hard X-ray spectrum in SS433 detected by INTEGRAL (Cherepashchuk et al. 2003) is $\gamma \sim 10$.

(iii) The fragmentation of the jets into dense clumps via the development of thermal instabilities has been studied (Bodo et al. 1985; Brinkmann et al. 1988). Bodo et al. (1985) found that, when it leaves the funnel ($r_{max} \lesssim 10^{11}$ cm), the gas begins to cool at distances $10^{11} \lesssim r \lesssim 10^{12}$ cm. On short time scales much less than the time scale for motion of the flow r/V_j, the thermal instabilities lead to the formation of a two-phase medium: cool, dense fragments ($T \approx 10^4$ K, $n \gtrsim 10^{16}$ cm^{-3}) embedded in a hot medium ($T \gtrsim 10^7$ K, $n \approx 10^{14}$ cm^{-3}). The dimensions of the condensations formed as a result of the instabilities are in the region of $l \approx 10^7$–10^8 cm, and are bounded from below by the thermal conductivity and from above by the condition that a pressure equilibrium should be established over the time for the development of the instability. Note that these calculations were undertaken before the discovery of the X-ray jets of SS433 and the associated measurement of the parameters of the jet gas, and they could be refined on the basis of the information now available. The resulting scales for the clouds are very close to those derived from analysis of the Balmer decrements of the jets (Panferov and Fabrika 1997), $l \sim 10^8$ cm. Recall also that, according to the data of Marshall et al. (2002), the temperature in the X-ray jets varies from $1.1 \cdot 10^8$ K to $6 \cdot 10^6$ K from their bases to their ends (the accuracy of the end values is limited by the weakness of the corresponding spectrum and blending of lines at soft energies), and the electron density similarly falls from $2 \cdot 10^{15}$ to $4 \cdot 10^{13}$ cm^{-3}.

In connection with the X-ray jets of SS433, Brinkmann et al. (1988) performed a numerical study of the evolution of a blob of material arising from initial density fluctuation in a uniform, hot gas forming a conical jet. The jet gas is in ionisation equilibrium via collisions and cools due to radiative losses and expansion of the jets. In the linear stage of the thermal instability, the blobs reach sizes similar to those obtained by Bodo et al. (1985) based on a linear analysis. However, in the subsequent non-linear stage, depending on a number of conditions, the blobs (clouds) can experience both catastrophic compression and expansion. In this case, a simple stationary state is not reached for any conditions, and the formation of clouds must be considered as a dynamical phenomenon in the evolution of the jets. It is certain that gas clouds with density contrasts $\gtrsim 10^3$ form in the cooling X-ray jets, and an additional mechanism to heat the gas in the clouds is required in order to maintain them at temperatures $\sim 10^4$ K over times $\sim 10^3 - 10^4$ s and to prevent collapse of the clouds. This additional heating could be radiative (by the collimated radiation of the funnel) or associated with the dissipation of shocks in the jet, radiative heating by radiation of these shocks (Begelman et al. 1980; Fabian and Rees 1979), or interaction of the precessing jets with the surrounding medium (Davidson and McCray 1980).

Depending on the magnitude of this heating in the X-ray jet, there could be substantial variations in estimates of both the physical conditions in the gas and of the kinetic luminosity of the jets. In addition, the resulting jet parameters depend strongly on assumptions about the jet structure. In their model of the X-ray jet, Brinkmann and Kawai (2000) took into account non-equilibrium effects and photoionisation of the gas, as well as the non-uniform density distribution across the jet. In particular, in their model with a narrow X-ray jet in which the density falls off with distance from the jet axis in accordance with a Gaussian distribution ($\theta_j = 1°\!\!.4$) that is in best

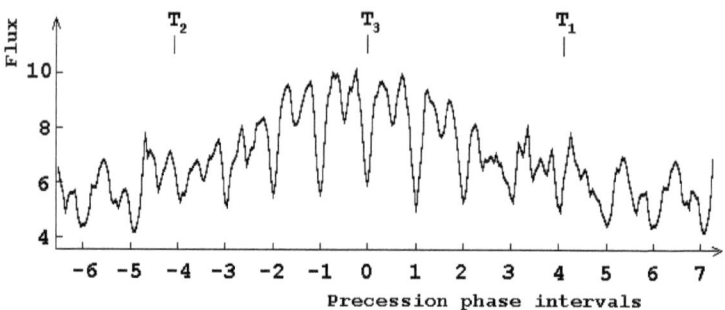

Figure 15. Flux density of SS433 in the V filtre (mJy) over the precessional cycle (Goranskii 2002). The precessional light curve is composed of mean orbital light curves, and is presented in terms of fractions of the orbital period.

agreement with the observations, the estimated kinetic luminosity is $L_k \sim 6 \cdot 10^{39}$ erg/s. This jet has a rarefied, hot atmosphere.

The Supercritical Accretion Disk and the Components from Photometric Data

The Light Curve of SS433: Precessional, Orbital and Nutational Variability

Extensive photometric observations have been obtained, and the results of such observations can be found in various reviews (Margon 1984; Cherepashchuk 1989, 2002). However, we will be interested primarily in data that can be used to draw conclusions about the SS433 accretion disk, or, more precisely, about the shape and structure of the gaseous envelope that surrounds the relativistic object. Much information about the structure of the disk and gaseous flows has been obtained from spectral studies. Accordingly, we will describe the available spectral data in more detail in the following section.

The light curve of SS433 is determined by three well-known periods – the precessional, orbital and nutational periods – as well as by sporadic variations – small-scale fluctuations, flares (described in the section "The Radio Jets and W50"), and active periods (Leibowitz et al. 1984; Kodaira et al. 1985; Kemp et al. 1986; Gladyshev et al. 1987; Mazeh et al. 1987; Goranskii et al. 1987, 1997, 1998ab; Zwitter et al. 1991; Fukue et al. 1997; Panferov et al. 1997; Irsmambetova 1997, 2001). Figure 15 shows the behaviour of the V flux density over a precessional cycle based on the data of Goranskii et al. (1998b). These data were translated into fluxes, averaged, and kindly presented to us by V. P. Goranskii (2002) especially for this review. The precessional light curve is shown in fractions of the orbital period; i.e., it is comprised of mean orbital curves obtained at various precessional phases over all the years in which photometric studies of SS433 have been undertaken. The orbital epochs when the accretion disk is eclipsed and precessional epochs when the disk is maximally turned toward the observer were combined. In all, 2400 individual measurements were used. When the accretion disk turns to face the observer during its precessional rotation ($\psi = 0$, time T_3), SS433 becomes brighter, the light curve becomes more regular, and eclipses become deeper and more well-defined. The depth of eclipses grows toward the blue, and appreciably decreases in the red and infrared.

At precessional phases when the disk is viewed edge-on (Fig. 15) and the jets lie in the plane of the sky, the light curve is very irregular, the primary eclipses of the accretion disk by the star (Min I, $\phi = 0$) become shallow, and eclipses of the star by the disk (Min II, $\phi = 0.5$) are sometimes difficult to distinguish. At these precessional phases, the light curve is very strongly distorted by flares (Goranskii et al. 1998b). Away from eclipses in quadratures, the precessional variability amplitude in the V band is $\approx 0\overset{m}{.}6$. At the centers of eclipses of the accretion disk, the precession amplitude is lower, $0\overset{m}{.}2$–$0\overset{m}{.}3$ (but is reliably determined); SS433 also becomes brighter when the disk is turned toward the observer and weaker when the disk is viewed edge-on. This led Goranskii et al. (1998b) to conclude that, independent of the precessional orientation, total eclipses of the disk in SS433 are never observed. The B–V colour varies little with the phases of the known periodic variations. When the brightness decreases at precessional phases when the disk is viewed edge-on or in eclipses, the object's V–R colour reddens – the hot source is eclipsed, while the red component of the emission (probably free–free radiation of the gas surrounding the system) is not. Colour variations are primarily observed during flares.

Such brightness variations with precessional phase could be associated with (i) simple variations of the orientation of a geometrically flat source, (ii) variations in the visibility of "hot spots" or funnels in the places where the jets emerge in the photosphere of the outflowing wind (Lipunov and Shakura 1982), or (iii) variations in the visibility of hot cocoons surrounding the bases of the jets. In the last two cases, the "disk" could be a disk-like body whose thickness grows with distance from the center and whose surface temperature grows toward the center. For example, roughly this shape has been assigned to the accretion disk in some computations of the SS433 light curves (Antokhina et al. 1992; Hirai and Fukue 2001). It is not ruled out that the "disk" plays purely a screening function. The outer edges of the disk can screen inner hot regions during the precession. In addition, it is completely unclear whether the shape of this body (which we traditionally call the disk) is determined by Keplerian rotation of the disk material. It is more likely that the shape is determined by the density, velocity and geometry of the wind, i.e., by the structure of the wind photosphere. This means that the size of the bright body surrounding the relativistic star could bear no relationship to the size of the Roche lobe of this star (see below), which gives rise to serious limitations when estimating the component mass ratio via modeling of the light curves.

Figure 16 shows orbital light curves of SS433 in the V band (Panferov et al. 1997) obtained from all published data by averaging the brightness measurements in intervals of the precessional phase. The flux $F_V = 1$ corresponds to $V = 14^m0$. The depth of the primary eclipses Min I ($\phi = 0$) varies appreciably as a function of the disk orientation; even their position varies. If the orbital and precessional rotations have opposite directions, as has been concluded in numerous studies and as should be true in the case of driven precession of the optical star (a slaved disk), the primary eclipses should lag the ephemerides before time T_3 ($\psi < 0$) and lead the ephemerides after time T_3 ($\psi > 0$) in models with a bright source coincident with the base of the approaching jet (models ii and iii above; Fabrika 1984; Gladyshev et al. 1987).

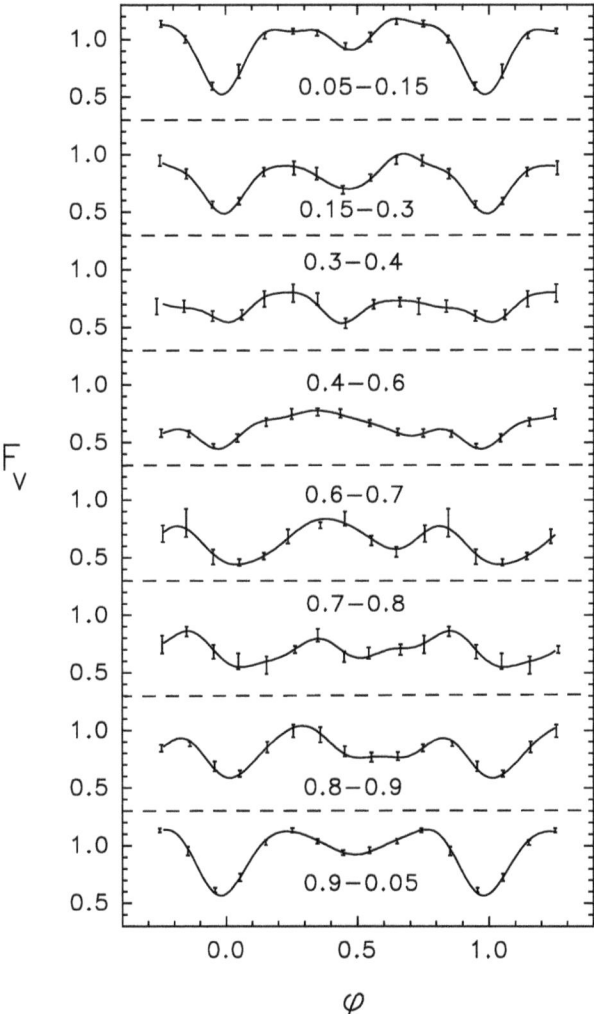

Figure 16. Orbital light curves of SS433 in the V filtre averaged over intervals of the precessional phase (Panferov et al. 1997).

The orbital light curves near the secondary eclipses Min II ($\phi \approx 0.5$) are even more variable. If the effect of reflection of the disk radiation from the surface of the donor star or from densified regions in the wind near the donor surface are sufficiently strong, then (again assuming the directions of the orbital and precessional rotations are opposite), at precessional phases before time T_3, the brightness should be higher at elongation $\phi = 0.2$–0.3 than at elongation $\phi = 0.7$–0.8, and Min II should lag somewhat behind the computed value. At precessional phases after time T_3, on the contrary, the brightness should be higher at elongation $\phi = 0.7$–0.8 and Min II should lead the computed value. This is approximately the pattern observed in Fig. 16. In addition, during the effect of heating, the maximum of the precessional brightness measured separately at the phases of the two elongations should also be shifted, due to the different conditions for the heating and the visibility of the star from the disk surface. More precisely, the precessional maximum occurs earlier than the ephemerides time T_3 at elongation $\phi = 0.2$–0.3, and occurs later than T_3 at elongation $\phi = 0.7$–0.8. This effect was noted by Gladyshev et al. (1987).

The presence of eclipses Min II in the X-ray (Gies et al. 2002a) might be associated with the reflection of light from the donor star, but could also represent eclipses of a region near the optical star if a modest fraction of the X-ray radiation is formed there in shocks in the colliding winds (Cherepashchuk et al. 1995). It seems reasonable to suppose that the effect of heating of the donor surface (or of perturbations in the wind) plays an important role in the brightness variability of SS433, but modeling of the light curves must be carried out to conclusively demonstrate that this is the case.

Apart from heating, two more effects that could distort the regular orbital variability are known. The first of these is nutational variations of the brightness, which distort the orbital variability in different ways at different precessional phases (Goranskii et al. 1998b). The second effect is absorption in the gas flowing out from the Lagrange point L2 behind the disk (Zwitter et al. 1991; Fabrika 1993). We will discuss these effects below.

The third reliably established period in the optical brightness variations is the nutational period of 6.28 days – the period with which nodding of the jets is observed. The total amplitude of the nutational brightness variations is $\Delta V = 0\overset{m}{.}17$ (Mazeh et al. 1981, 1987; Goranskii et al. 1998b). The additional source of red light (in the R band) does not show nutational variations; i.e., the amplitude of the variations of the blue source must be slightly higher than in the V band.

Here, we present ephemerides of the photometric variability with the periods of the precession P_{pr} (time of brightness maximum T_3), orbital motion P_{orb} (the center of the eclipse of the accretion disk Min I) and nutation P_{nut} (maximum brightness in the V band) based on the data of Goranskii et al. (1998b), which are the most recently published. These ephemerides may prove useful in computations and planning of observations.

$$T_3 = \text{JD } 2450000 + 162\overset{d}{.}15 \cdot E$$

$$\text{Min I} = \text{JD } 2450023\overset{d}{.}62 \pm 0\overset{d}{.}26 + (13\overset{d}{.}08211 \pm 0\overset{d}{.}00008) \cdot E$$

$$\text{Max} = \text{JD } 2450000\overset{d}{.}94 + 6\overset{d}{.}2877 \cdot E$$

The orbital elements are the most accurate at the end of 2001, and were confirmed after publication (Goranskii et al. 1998b) by recent observations of eclipses (Goranskii 2002). When the elements of the precessional period are compared with the spectral precessional elements presented above (Eikenberry et al. 2001; "The Optical Jets"), they are in reasonable agreement given at the instabilities and inaccuracies that are usual for precessional clocks. The accuracy of the spectral precessional period should naturally be higher than that of the photometric precessional period, since the amplitude of the shifts of the moving lines is much larger than the uncertainties in the positions of lines in the spectrum (by a factor of $\sim 10^3$). In photometric observations, the amplitude of the precessional variability does not exceed the typical measurement errors by such a large factor. Nevertheless, we study the precession of the accretion disk via photometry, while we observe the motion of the jets via spectroscopy, so that a comparison of the precessional motions of the jets and disk may have fundamental value.

The Light Curve in Active and Quiescent States

The light curve in Fig. 15 was obtained by averaging all data in both active and quiescent states of the object. An analysis of the precessional and orbital variability separately in active and quiescent states was carried out by Fabrika and Irsmambetova (2002), using the same photometric database of the Sternberg Astronomical Institute (1979–1996). Active states of SS433 (see the Introduction and the section "The Radio Jets and W50") were identified using the results of the Green Bank Interferometer monitoring program (http://www.gb.nrao.edu/fgdoss/gbi/gbint.html), as well as on the basis of the optical data themselves. The precessional light curves in active and quiescent states are appreciably different. In quiescent states, the mean brightness of SS433 (of the accretion disk) outside of eclipse depends strongly on the precessional phase. The object becomes weaker in the V band by approximately a factor of 2.2 when the accretion disk turns from being maximally turned to face the observer to being viewed edge-on. However, in active states (excluding obvious flares), the brightness of SS433 depends only very weakly on the disk orientation. It is especially important that, when the disk is turned to maximally face the observer (time T_3), the brightnesses in active and quiescent states are the same. When the disk is viewed edge-on (times T_1, T_2), the object is significantly weaker in quiescent states than in active states.

The orbital light curves in the active and quiescent states are approximately the same – primary and secondary eclipses are also observed. At the center of eclipses of the accretion disk by the star (primary minima, $\phi \approx 0$), SS433 brightens by a factor of ≈ 1.6 in active states compared to quiescent states. The precessional modulation at the center of these eclipses is roughly the same in active and quiescent states: the brightness is slightly higher when the disk is turned to face the observer than when it is edge-on. There exists some source of radiation in SS433 that is not eclipsed by the star even in blue light. This may be associated with radiation scattered in the outer gas of the wind.

The photometric behaviour of SS433 in active and quiescent states suggests a model (Fabrika and Irsmambetova 2002), in which the optical brightness of the object is primarily determined by some hot body in the central part of the accretion disk, which could be two hot gaseous cocoons surrounding the inner (X-ray) jets. In active states, the size of these cocoons increases, and they are not blocked from the observer by the rim of the accretion disk when it is viewed edge-on, or by the donor star during primary eclipses. The approximate equality of the amplitudes of the precessional modulation and of the primary eclipses indicate that the size of the donor is approximately equal to the size of the outer rim of the disk in projection onto the plane of the sky.

When the disk is maximally facing the observer, the mean brightness at elongations is the same in active and quiescent states. This may indicate that the cocoon surrounding the approaching jet is not blocked by the outer rim of the disk at precessional phases when the disk is maximally facing the observer, so that its radiation can be observed right down to its base, and that the luminosity of the cocoon does not depend appreciably on its size (or the activity state of SS433). This may be the case if the cocoon gas scatters radiation ($\tau_T \sim 1$) arriving from inner regions – from a funnel in the accretion disk or the wind. These cocoons may be identified with a source of UV radiation in which Dolan *et al.* (1997) detected strong linear polarization directed along the jets, or with the source of the two-peaked He II $\lambda 4686$ line in the spectrum of SS433 (see next section).

The Nutational Clock and Time for the Passage of Material through the Disk

Due to the conjugacy of the nutational period of 6.28 days with the orbital and precessional periods, the distortions of the orbital light curves due to the nutational oscillations will depend on the precessional phase (Goranskii *et al.* 1998b). In particular, as can be found from the ephemerides presented above, the nutational maxima at time T_3 occur at elongations. The nutational variability leads to some shifts in the positions of eclipses. The magnitude and sign of these shifts depend on the precessional phase. Model fitting of the orbital light curves of SS433 should take these nutational oscillations into account.

Information about the phases of the spectral and photometric variability with the precessional and nutational periods is very important for our understanding of the precession mechanism operating in SS433. As was already mentioned, the most successful scenario for the precession is driven precession of the donor star, whose rotational axis is not coincident with the orbital axis, with a drifting or "slaved" accretion disk (Shakura 1972; Roberts 1974; van den Heuvel et al. 1980; Whitmire and Matese 1980; Hut and van den Heuvel 1981; Matese and Whitmire 1982). The massive donor star can undergo driven precession (Papaloizou and Pringle 1982; Collins 1985). In their analysis of periodic perturbations of the disk, which does not lie in the orbital plane, by the gravitational field of the donor star, Katz et al. (1982) concluded that the precessional and nutational motions of SS433 can best be explained by a slaved-disk model. The analysis of nutational synodic phenomena is an accurate tool, and the amplitudes of these motions and possible variations of the period depend directly on many parameters of the binary system. Collins (1985), Collins and Newsom (1986) and Collins and Newsom (1988) developed a dynamical model for SS433 (see also the revised dynamical model in the recent paper by Collins and Scher (2002)) in which it was possible to deduce the properties of the precessing star, orbital eccentricity, and apsidal motions.

In terms of frequencies, the nutational period of 6.28 days looks like $f_{nut} = 2f_{orb} + f_{pr}$. In practice, the perturbations of the accretion disk occur with a period of 6.06 days; this is the nodding of the disk motion (Katz et al. 1982), $f_{nod} = 2f_{orb} + 2f_{pr}$. As could be expected ("The Radio Jets and W50"), flares in the SS433 system should "feel" precisely this nodding period; i.e., the beating period in a coordinate system rotating with the accretion disk or the star. However, in the observer's system, where photometric and spectral effects also depend on the angle between the line of sight and the disk or jet axis, this period becomes 6.28 days.

Perturbations of the disk (or accretion flow) of SS433 lead to corresponding variations in the jet inclination if the time for the material to pass through the disk is not too long, and does not greatly exceed the perturbation period. Information about variations of the inclination of the rotational moment of the outer parts of the disk should reach the inner regions (the source of the jets) without being appreciably distorted or smoothed. We will set aside questions concerning the structure of the inclined disk and the interactions of various harmonics in the perturbations of the outer edge of the disk produced by the gravitational field of the donor star (Katz et al. 1982).

Qualitatively, the model is such that, at times of elongations, the perturbation of the disk by the donor star leads to a shift of the disk rotational axis in the plane of the sky (further, this perturbation affects the direction of propagation of the jets), but this does not change the inclination of the jets to the line of sight. Therefore, perturbations applied to the disk at times of elongations do not lead to shifts in the moving lines. Perturbations of the edge of the disk at times of conjunctions are directed perpendicular to the plane of the sky, and will therefore change the inclination of the jets. In particular, at precessional phases close to T_3 (the inclination of the jets and of the disk axis to the line of sight is $\approx 60°$), perturbations at conjunctions tend to align the disk in the orbital plane. Since the inclination of the orbital axis to the line of sight is $\approx 78°$, the effect is to cause the $H\alpha^{\pm}$ lines to approach each other, while the brightness of the system weakens. A quarter orbital period later, at elongation, the disk and jets return to their initial positions, the $H\alpha^{\pm}$ lines move apart, and the brightness grows. Of course, we will see the reaction of the disk and jets to the gravitational perturbations by the star only after the time required for the material to pass through the disk and for the jets to move to the region of efficient line radiation. In reality, in addition to perturbations of the outer parts of the disk, we should bear in mind that the conditions for the formation of the disk (the locations of the relativistic star relative to the equator of the donor) change with the nutational phase. The heating of the donor surface is inhomogeneous due to shadowing by the disk and flows. All these effects change the geometry for the mass transport, and depend on the phase of the beating between the precessional and orbital periods.

Mazeh et al. (1987) found that the jet nutational phases derived from spectral and photometric data are not quite coincident: the photometric maxima lead the nutational radial-velocity extrema by about one day. Goranskii et al. (1998b) confirmed this result by showing based on 16 years of data (950 nutational periods) that the nutational increase in the brightness of SS433 coincides with the maximum shifts of the $H\alpha^-$ lines toward the blue, with a small but significant phase shift $\Delta\phi_{nut} = 0.10 \pm 0.02$ (Goranskii et al., 1998b). The nutational deviations of the jets lag the optical variability. This phase shift corresponds to a delay of 0.6 days, or a distance travelled by the jets of $\approx 4 \cdot 10^{14}$ cm. This is precisely the distance at which the maximum radiation of the jets in the moving $H\alpha$ lines is achieved (Fabrika and Borisov 1987; Vermeulen et al. 1993a). This leads us to conclude that the nutational oscillations of the brightness are associated with the bases of the jets. There are also nutational variations in the X-ray (Gies et al. 2002a), with the phase of the maximum X-ray flux roughly coinciding with the phase of the maximum in the optical.

However, the energetics of the jets are insufficient to explain the 6-day variations in the optical. It is natural to propose that the entire central engine of the disk–jets system participates in the nutational variability. The amplitude of these variations is about $\sim 10^{39}$ erg/s, comparable to the kinetic luminosity of the jets. Even the X-ray luminosity of the jets is roughly a factor of $\sim 10^4$ lower than the bolometric luminosity of the accretion disk, and the optical continuum luminosity of the jets should be lower than the X-ray luminosity, so that the optical radiation of the jets cannot explain the nutational brightness variations ($\Delta V \approx 0\overset{m}{.}17$). These variations may occur in inner regions where the bases of the jets are located and which, like the jets, should execute precessional and nutational motions.

If the brightness variations with the nutational period are caused by nodding of the outer parts of the disk or the donor star, the time for the passage of the material across the disk is either approximately "zero" (e.g., equal to the free-fall time, which is several tenths of a day), or is equal to a multiple of the nutational period. However, the problem of the short implied time for the passage of material across the SS433 disk arises for other reasons as well, that are not associated with photometric variability.

Using the model of Katz et al. (1982) for the nodding motions of the accretion disk, Gies et al. (2002a) found based on new observations of the moving jet lines that the nutational deviations of the jets occur 1.0 day later than the primary perturbations at elongations (or at conjunctions; the important thing here is the fixing of the time in terms of the orbital clock). In this way, Gies et al. (2002a) confirm the delay of oscillations of the jet relative to the photometric variations. While 0.6 days (the delay of the jet nutation relative to the photometric variability found by Goranskii et al. (1998b)) agrees well with the required time for the motion of the jet gas to the region of Hα radiation at a distance $R_j \approx 4 \cdot 10^{14}$ cm, it seems that the remaining time of 0.4 days must be the time for the passage of the signal across the accretion disk? It is certain that such analyses can provide information about the structure of the SS433 disk. If the result of Gies et al. (2002a) indicating a time delay of 1.0 day is confirmed by future observations (or by new analyses of existing data using more accurate, modern ephemerides), this may mean that the material in the accretion disk reaches inner regions over the free-fall time. The tilted disk could consist, for example, of flows of material that lose momentum in shock waves.

In fact, a time delay of about one day was found in early studies, before the nutational photometric variability had been discovered. Collins and Newsom (1986) found in the framework of a dynamical model that the jets deviate $0\overset{d}{.}83 \pm 0\overset{d}{.}2$ after the times of perturbation of the precessing body. They proposed that the time for the passage of material across the disk and further to the region of optical jet emission was $0\overset{d}{.}8$ plus a time equal to a multiple of the nutational period. Katz et al. (1982) also found a time delay of $0\overset{d}{.}9$ in a model for the nodding motion of the disk (the delay of the deviation of the jets after the perturbations at elongations), but they used data on the radial velocities of the He II λ4686 line obtained by Crampton and Hutchings (1981a) to determine the orbital phase. Thus, the delay of the times of noddings of the jets relative to the times of perturbations of the accretion disk of ≈ 1 day can be considered to be reasonably well established.

One test of the possibility that the time for the passage of material across the disk is a multiple of the nutational period would be a comparison of the precessional phases determined photometrically and spectroscopically. It is thought that the known precessional brightness variability is associated with precession of the disk. However, the precessional brightness fluctuations (Gladyshev et al. 1987; Goranskii et al. 1998b) do not lead the precessional motions of the jets, as they should; in contrast, a delay of the precessional photometric wave relative to the jet precession is observed. This delay comprises no more than three to four days (Goranskii et al. 1998b), and could easily be associated with inaccuracy of the photometric ephemerides for the precession. Nevertheless, the approximate phase coincidence of the photometric and spectral precessional periods does not support the idea that the time for the passage of material in the disk is a multiple of the nutational period. However, it is not possible to rule out this idea if the precessional variability is in no way related to precession of outer parts of the accretion disk.

In summary, we can conclude that the nutational brightness variations are associated with variations in the orientation of hot inner regions in the places where the jets emerge (for example, cocoons of hot gas). The precessional variability most likely has the same origin. Independent of the specific interpretation of the photometric variability, the problem exists of the small time inferred for the passage of material across the disk, which is close to the free-fall time. However, we cannot rule out the possibility that the time for the passage of material across the disk is a multiple of the nutational period.

Outflows in the Disk Plane and Gaseous Flows

As soon as the disk is oriented edge-on (beginning with precessional phase ≈ 0.3), the gas trail beyond the disk can cross the line of sight at orbital phases $\phi > 0.5$. Absorption in this flow of gas lost by the system could plausibly give rise to an appreciable dimming of the brightness, roughly as is observed in Fig. 16, where we see a significant weakening of the brightness after Min II at precessional phases $\psi = 0.3-0.7$. Absorption in such a flow could also be significant even when the accretion disk is not edge-on. Due to the precessional shift of the point L2, the orbital phase where we could expect absorption will also shift, and this will shift the position of Min II in approximately the same direction as in the case of heating of the surface of the optical star. It is not possible to determine which of these effects distorts the light curve more strongly without specialised modeling of the precessional and orbital brightness modulations.

If absorption in flows of gas lost by the system is indeed that substantial, the optical depth of the outflowing material to Thomson scattering will be ~ 1, and the radial density will be $N_H \sim 10^{24}$ cm^{-3}. In addition to its influence on the optical light curves (Zwitter et al. 1991; Fabrika 1993), it is believed that precisely this material absorbs the radiation of the receding X-ray jet (Kotani et al. 1996) and gives rise to the radiation of the equatorial VLBI disk further from the system (Paragi et al. 1999, 2000; Blundell et al. 2001), and may even be observed on larger scales (seconds–minutes of arc) in optical line emission (Fabrika 1993). There is direct evidence for the presence of spreading flows of material in the plane of the accretion disk, which we will describe when we discuss the spectra of SS433.

The depth and shape of X-ray eclipses of the accretion disk vary substantially, depending on the precessional phase (Kawai et al. 1989; Brinkmann et al. 1991; Yuan et al. 1995; Kotani et al. 1997b). Some of these X-ray eclipses are in reasonable agreement with optical eclipses. Thanks to the ability to obtain nearly continuous observations, observations of X-ray eclipses make it possible to directly study the structure of the accretion disk, and to map this region and its near environment.

In addition to eclipses by the optical star, other substantial decreases in the X-ray brightness have been detected (with amplitudes as large as those during the eclipses, or even larger), which it has not been possible to reconcile with the canonical picture of eclipses in the binary system (Brinkmann et al. 1989). These "unforeseen" eclipses in SS433 are not fully understood, and these brightness dips are most likely due to absorption in both accretion flows in the system and flows of gas out of the system roughly in the plane of the accretion disk. The orbital phases and structure of these brightness decreases are consistent with them being associated with a turning point of a flow near the Roche lobe (Brinkmann et al. 1991; Lubow and Shu 1975). Absorption could also arise in gaseous clouds above the plane of the disk and in gaseous outflows from the system. Hydrodynamical computations of the formation of the accretion disk when the donor overfills its critical Roche lobe (Sawada et al. 1986; Chakrabarti and Matsuda 1992) show that a complex structure with spiral shocks should exist in the disk, with powerful outflows of gas beyond the disk. Computations of the formation of an inclined slaved accretion disk (Bisikalo et al. 1999) also indicate a very complex structure for the gaseous clouds outside the plane of the disk and the presence of a shock wave along the stream of accreting material.

Sharp Brightness Decreases

The sharp and deep dips in the optical brightness of the object that are observed at various precessional and orbital phases (Henson et al. 1982; Kemp et al. 1986; Gladyshev et al. 1987; Goranskii et al. 1998b) may prove important for our understanding of the nature of the components of SS433. The deepest observed dip, to V= $17^m\!.3$, occurred during a primary eclipse when the disk was oriented edge-on (Henson et al. 1982); the brightness of SS433 dropped by $2^m\!.5$ compared to its normal level for the corresponding orbital and precessional phases. It is interesting that the object was in an active state at that time, and its brightness was high $0^d\!.5$ before and $0^d\!.5$ after this record brightness dip (Goranskii et al. 1998b). Other brightness decreases by $1^m\!.9$ and $1^m\!.1$ relative to the normal brightness level have been observed outside of eclipse. These dips imply that the optical star in SS433 is at least a factor of 20 weaker than the accretion disk in the V band, since the usual brightness of SS433 is V = $14^m\!.0$. Adopting a distance to the object of 5.0 kpc and an absorption of $A_V = 8^m\!.0$, we find that the luminosity of the donor is $M_V > -4^m\!.5$, if, of course, this star itself is not subject to eclipses during these brightness dips.

These sharp decreases in the disk radiation are surprising, all the more so because the strongest dip occurred in an active state. Is it possible that the mass-transfer process is sufficiently non-stationary during times of activity that material does not reach the disk for some time? It may be that, as in the case of the "unforeseen" X-ray eclipses, we must think about eclipses of an optical source in the accretion disk by gaseous flows. An alternative scenario in which the V brightness dips are associated, on the contrary, with sharp increases in the rate with which material is supplied to the disk is also possible. Roughly speaking, the luminosity of the supercritical disk does not depend on the accretion rate $\dot{M}_0$. The size of the photosphere of the supercritical disk's wind is $R_{ph} \propto \dot{M}_0$, and the temperature of the photosphere is $T_{ph} \propto \dot{M}_0^{-2}$. When there is a sharp increase in $\dot{M}_0$, for example by a factor of ~ 10, the photosphere temperature falls to several thousand Kelvin, the size of the wind photosphere grows by a factor of several tens (or even more, since the main source of opacity will become free–free transitions and molecules rather than Thomson scattering), and the entire binary system will end up being deep beneath the wind photosphere.

Observations of the sharp brightness dips in other spectral regions (or at least in two different optical bands) would help elucidate the nature of these strange brightness decreases. Note that such dips argue in favour of a short time for the passage of material across the disk.

The Spectral Energy Distribution and Parameters of the Components

Outside of eclipses and at precessional phases when the accretion disk is turned to face the observer, the luminosity and temperature of the object are substantially enhanced. It is difficult to accurately estimate the temperature of the radiation from optical photometric data, since it is necessary to work in the far Jeans region of the spectrum. Based on $WBVR$ photometry, Cherepashchuk et al. (1982) found that the spectrum of SS433 at brightness maximum was consistent with that of a black body with a radiation temperature of $T \gtrsim 50\,000$ K subject to absorption $A_V = 7^m\!\!.4 - 8^m\!\!.3$, with the hot body having a radius $R \approx 2 \cdot 10^{12}$ cm and bolometric luminosity $L_{bol} \gtrsim 10^{40}$ erg/s. Murdin et al. (1980) arrived at essentially the same conclusion in one of the earliest studies of SS433 ($A_V \approx 8^m$, $T \sim 30\,000$ K, $L_{bol} \sim 3 \cdot 10^{39}$ erg/s). Wagner (1986) drew similar conclusions based on spectrophotometric data ($A_V = 7^m\!\!.8 \pm 0^m\!\!.5$, $T \sim 32\,500$ K, $R \sim 30 R_\odot$, $L_{bol} \sim 4.4 \cdot 10^{39}$ erg/s), and confirmed that the source becomes hotter when the bright, precessing body is observed nearer to the pole. Formally, in the presence of absorption $A_V > 8^m\!\!.2$, the temperature of the source derived from optical photometric data approaches infinity.

Observations with the HST/HSP confirmed the high temperature of the radiation of SS433 derived from optical photometry. Figure 17 presents the observed fluxes in the F227M and $UBVR$ bands based on the data of Dolan et al. (1997). The ultraviolet observations were conducted in a band centered on 2270 Å during both bright (precessional phase close to zero and orbital phase close to elongation) and

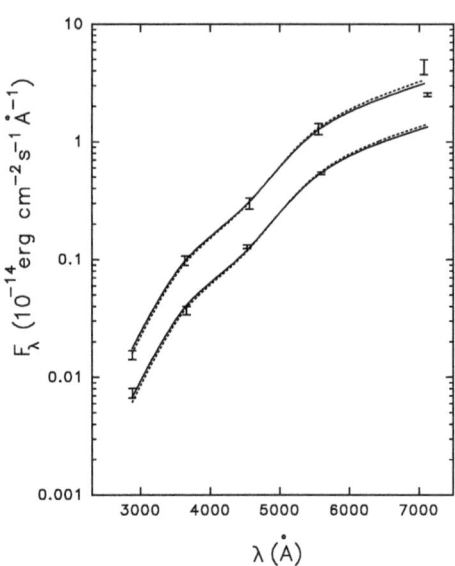

Figure 17. Observed fluxes of SS433 in the F227M and $UBVR$ bands (Dolan et al. 1997). The upper panel shows the bright precessional state and the phases of orbital elongations, and the lower panel shows the edge-on precessional state and the phase of eclipse of the accretion disk. Approximations using a single blackbody spectrum taking into account interstellar absorption at the isophotal wavelengths of the SS433 spectrum are shown. The solid curves show the best solution for the UBV data only, while the dotted curves show the solution for UBV + F227M. The intensity of the additional radiation in the R band is independent of the precessional and orbital phases, and is equal to $\Delta F_R = 1.2 \cdot 10^{-14}\,\mathrm{erg/cm^2\,s\,Å}$.

dim (edge-on precessional orientation and phases of eclipses of the accretion disk) states. The mean $UBVR$ fluxes for these two states were derived from the optical photometry database. Approximations using a single blackbody spectrum (without using the R data) show that the temperature of the object's radiation in the bright state is $T = 72\,000 \pm 20\,000$, with absorption $A_V = 8.4\,(+0.2, -0.3)$, source radius $R = 2.0 \cdot 10^{12}$ cm and bolometric luminosity $L_{bol} = (4\text{–}8) \cdot 10^{40}$ erg/s. It was not possible to obtain a satisfactory approximation to the spectrum using a single blackbody emitter in the dim state; Fig. 17 shows the spectrum for $T = 4.9\,(+2.9\text{--}1.5) \cdot 10^4$ K, $A_V = 8.2 \pm 0.5$, $R = 1.5 \cdot 10^{12}$ cm and $L_{bol} \approx 1 \cdot 10^{40}$ erg/s.

These data also confirm the presence of a cool envelope, as follows from optical and infrared photometry. There is an additional source of radiation in the R band (Fig. 17), independent of the precessional and orbital phases, equal to $\Delta F_R = 1.2 \cdot 10^{-14}$ erg/cm s Å. This additional "red light" has now also been distinguished in analyses of flares of SS433 (Irsmambetova 1997, 2000; Goranskii et al. 1998a). As was already mentioned above ("The Radio Jets and W50"), SS433 displays both "white" (i.e., hot) and "red" optical flares. Red flares are due to the excitation of free–free emitting gas located around the system, and show delays in the R and I bands by several hours relative to the corresponding flares in the V band. In addition, as a rule, SS433 reddens in the transition to the active state.

Observations in the far UV were conducted by Gies et al. (2002a) using the HST/STIS at 1150–1700 Å. Unfortunately, SS433 was not detected in these observations, and it was only possible to place an upper limit on the flux. The precessional phase at the observing epoch corresponded to the dim state (edge-on). Using these upper limits, the data of Dolan et al. (1997), archival IUE observations and the optical spectrophotometric data of Wagner (1986), Gies et al. (2002a) found that the temperature of the source derived by Wagner (1986) based on optical data in the dim state ($T = 21\,000$ K for $A_V = 7\overset{m}{.}8$) was consistent with the upper limits for the far-UV flux obtained from the STIS observations. The implied radius of the hot source was $R = 2.3 \cdot 10^{12}$ cm.

The uncertainties in estimates of the temperature of the accretion disk (more precisely, of the photosphere of the disk wind) depend on the photometric data used, assumptions about the absorption law, and the complexity of the source. It is usually assumed that this is a single object radiating as a blackbody. Note also that the spectral energy distribution of the supercritical disk should not correspond exactly to a blackbody distribution (Lipunova 1999; Okuda and Fujita 2000).

Fuchs et al. (2002) detected He I and He II emission in an infrared spectrum obtained with ISOPHOT (2.4–4.8 and 6–12 μm), and concluded that the spectrum was similar to a WN spectrum. However, it is known from the radial-velocity curves that the He I and He II lines in SS433 are radiated in gas flows, in the accretion disk and in the accretion disk wind rather than in the donor wind. The observed spectral energy distribution at 4–12 μm agrees well with that expected for free–free radiation in an optically thick envelope, with $\alpha = 0.6$ in the power-law spectrum $F_\nu \propto \nu^\alpha$. Optically thin free–free radiation is observed between 2.4 and 4 μm with $\alpha = -0.1$ (see also Giles et al. 1979; McAlary and McLaren 1980); and the spectrum is well described by blackbody radiation by dust with $T \sim 150$ K at the longer wavelengths, 12–60 μm. Fuchs et al. (2002) estimated the mass-loss rate in the wind to be $\dot{M}_w \approx 1.0 \cdot 10^{-4} \, M_\odot/\mathrm{yr}$.

Cherepashchuk et al. (1982) estimated the parameters of the optical star using four-band photometric data obtained during eclipses of the accretion disk. They found that the temperature of this star is in the range $T_s = 13\,000 - 43\,000$ K, while its radius is $R_s = 28 - 47 \, R_\odot$. No spectral lines from the star were detected. It was suggested (Fabrika 1998) that optical lines should be searched for among high-order terms of the Balmer series, since the Balmer emission decrement is very steep while the stellar absorption decrement is flat; however, blue spectra of SS433 with the required quality were not obtained.

Recently, Gies et al. (2002ab) presented promising data. They proposed that the best phases for searches for absorption lines of the star were during eclipses of the disk and precessional phases $\psi \approx 0$, when outflows of gas from the system do not cross the line of sight. They detected an He I $\lambda 6678$ absorption component at these phases, which moved over three successive nights in agreement with the orbital motion of the donor. Gies et al. (2002b) detected absorption lines in the blue, which are probably associated with the donor photosphere. We will describe these results in more detail in the next section.

Polarization of the Optical and UV Radiation

The results of linear polarization measurements could prove important for our understanding of the accretion disk of SS433. The optical polarization in the $BVRI$ bands is about 2% and is variable

(McLean and Tapia 1980; Efimov et al. 1984; Dolan et al. 1997), with the polarization angle directed along the plane of the accretion disk (perpendicular to the jets). This is in good agreement with the expectations for scattering of the disk radiation by free electrons located above the disk. The degree of polarization grows slightly toward the blue. However, Dolan et al. (1997) found that the character of the polarization changes sharply in the UV. The degree of polarization in the U band is $8.9 \pm 0.9\%$, and has already grown to $13.7 \pm 3.0\%$ in the 2770 Å band (FWHM = 340 Å, HST/HSP). In addition, the orientation of the polarization changes by $90°$, so that it is directed along the jets in the UV. The UV polarization is very variable from observation to observation, and can sometimes reach 20% in the 2770 Å band.

Dolan et al. (1997) proposed that the UV polarization arises due to Rayleigh scattering on neutral hydrogen atoms located in the plane of the accretion disk beyond the binary system. This imposes strong constraints on geometrical models for the sources of the optical and UV radiation in the SS433 accretion disk, since it remains unclear why the UV radiation efficiently scatters on a gas in this plane, while the optical radiation does not.

In the context of the most recent X-ray observations and optical spectral observations during eclipses (see below), a picture is sketched out for the inner regions of SS433, where the bases of the jets are shrouded in hot gas immediately above the disk. The temperature of this cocoon falls sharply from $\sim 10^8$ K on its axis (the region of the jet motions) to $\sim 5 \cdot 10^4$ K at the edges (the He II cocoon). The region of stationary X-ray Fe-line fluorescence is probably located there as well. The UV radiation also forms in inner regions, since the effects of eclipses by the optical star are visible in this radiation (and in its polarization). If the UV radiation forms in the gas surrounding the bases of the jets or in cocoons, this could lead to its polarization. The emerging radiation will be Thomson scattered by the cocoon gas, with the plane of polarization of the scattered radiation oriented along the jets, in agreement with the data of Dolan et al. (1997). A more detailed quantitative model for the region of UV radiation and computations of the polarization of the emerging radiation are very desirable.

The Supercritical Disk from Spectroscopic Data

The "Stationary" Spectrum of SS433

The "stationary" lines in the spectrum of SS433 include the strongest hydrogen lines, as well as emission lines of He I, He II, C III, N III, and weaker Fe II emission (Murdin et al. 1980; Crampton and Hutchings 1981ab; Dopita and Cherepashchuk 1981; Falomo et al. 1987; Filippenko et al. 1988; Kopylov et al. 1989; Fabrika et al. 1997a; Gies et al. 2002a; Fuchs et al. 2002). The spectrum is indeed reminiscent of the spectrum of a WR star, or more precisely, of a member of the recently identified class of late WN stars, of type WN10–WN11 (Crowther and Smith 1997; Bohannan and Crowther 1999). These stars, in turn, are close relatives of the Luminous Blue Variables (LBVs; Humphreys and Davidson 1994), which display evidence for hot outflowing atmospheres and late-WN spectra when their brightness is weak. The similarity between the spectrum of SS433 and WNL spectra is probably not a coincidence. The conditions for the formation of the SS433 spectrum and lines in the wind of the supercritical disk are probably close to those in the winds of late-WN stars, and the chemical compositions might also be similar if the donor in SS433 is a fairly evolved star.

It is primarily the variability and behaviour of the stationary $H\alpha$ that has been studied in detail, since it is the strongest line in the spectrum and is a convenient line for observational studies, since the object is very bright in the red part of the spectrum. The $H\alpha$ equivalent width is strongly variable. It is usually in the range 100–300 Å, but sometimes reaches 1000 Å. The profiles of powerful emission lines (FWHM $\sim 1\,000$ km/s, full width at the base $3\,000$–$5\,000$ km/s) vary with the precessional phase, becoming more structured and often broader when the accretion disk is oriented edge-on. In contrast, the intensity of the lines grows appreciably toward phases near T_3. For example, on average, the $H\alpha$ luminosity varies by a factor of three with the precessional phase (Asadullaev and Cherepashchuk 1986; Fabrika et al. 1997a), and is $L(H\alpha) \approx 7 \cdot 10^{36}$ erg/s when the disk is maximally turned to face the observer. The structure of the hydrogen-line profiles is determined by gas motions and has a large-scale nature; strong emission components appear, often in the blue and red wings (±1 000 km/s). It is not known, however, whether some of the profile structure arises due to line absorption from the blue (and red) side. The hydrogen and He I lines also show "regular" P Cygni profiles, which arise in the wind, at precessional phases near edge-on orientation of the disk. At such times, the Fe II lines can have such strong blue absorption components that the emission components disappear. This indicates strong non-isotropy of the wind. All the lines in the SS433 spectrum form either in various places in the wind flowing out from the disk or in gaseous flows in the system. The line profiles and fluxes vary strongly during flares.

He II Radial-velocity Curves and the Mass Function

Studies of the line radial velocities and the behaviour of the lines during eclipses of the accretion disk have provided much information about the gaseous flows in SS433 and the structure of the disk wind, both in inner regions where the bases of the jets are located and just beyond the limits of the binary system. The observational manifestations of the flows and wind vary strongly with the precessional and orbital phases. The inclination of the system's axis to the line of sight is well known from analyses based on the kinematic model ($i \approx 78°$), so that it is not an unknown parameter, as is usually the case with studies of binary stars. Here, we will describe investigations of the behaviour of the He II $\lambda 4686$ line, which yield, in particular, the mass function of SS433, and then describe studies of gaseous flows and the wind in the system.

The He II line possesses the highest excitation potential of all the lines in the SS433 spectrum, and is also the only line whose orbital motion reflects the motion of the relativistic star (Crampton and Hutchings 1981a). SS433 is probably the only star for which so many contradictory reports of its mass have been published. This is associated with objective difficulties. The mass ratio can be obtained from the duration of eclipses, but the size of the body surrounding the relativistic star turns out to depend on the precessional phase and the observing wavelength. In addition, the size of the wind photosphere is in no way related to the size of the Roche lobe of the relativistic component. The mass function can be derived from the orbital variability of the line radial velocities, but the line radiation occurs primarily in gas flows, so that the radial velocities vary appreciably with the precessional phase.

Crampton and Hutchings (1981a) found that the half amplitude of the He II λ4686 radial velocities is $K = 195\pm19$ km/s, with a mean velocity (the γ velocity) for the line of $V_0 = +27\pm13$ km/s, while the orbital phase of the transition of the radial velocity through the γ velocity from the positive to the negative region is (recalculated to the modern, precise ephemerides) $\phi_0 = -0.01 \pm 0.02$. This is the phase of the upper conjunction of the source, and coincides with the time of the upper conjunction of the accretion disk. The mass function derived from these data is $f(M) = M_o^3 / (M_x + M_o)^2 \approx 10.8\,M_\odot$, where M_o and M_x are the masses of the optical and relativistic star, respectively. The scatter of points about the radial-velocity curve is appreciable (as in later studies), and is not associated with measurement errors or insufficient spectral resolution, but with real variability of the radial velocity and the structure of the line profile. In particular, the neighbouring Hβ line is substantially brighter and displays a sharp one-peaked profile (showing that the measurement accuracy for this line is high), but Crampton and Hutchings (1981a) found that the orbital curve of the radial velocities of this line (and also of the He I and Fe II lines) varies substantially with the season. When it is exhibiting regular behaviour, the Hβ radial velocity varies with a half-amplitude of ≈ 80 km/s, and has a mean velocity of $\approx +220$ km/s and $\phi_0 = 0.25$. These last two values indicate that the Hβ radial velocity is appreciably distorted by effects such as absorption in the wind, and the line itself forms in a gaseous flow.

Fabrika and Bychkova (1990) found that the radial-velocity orbital curve of the He II line depends on the precessional phase: when the accretion disk is maximally turned to face the observer (T_3, $0.9 \lesssim \psi \lesssim 0.1$), the main contribution to the emission of this line is made by the accretion disk (more precisely, by regions near the relativistic object), while the main contribution at other precessional phases is made by a gaseous flow outside the accretion-disk region. We will describe the structure of this flow in more detail below. Near the phase of T_3, the He II radial-velocity curve is similar to that obtained by Crampton and Hutchings (1981a): $K = 175 \pm 20$ km/s, $V_0 = -13 \pm 12$ km/s, and $\phi_0 = 0.03 \pm 0.01$, corresponding to the mass function $f(M) \approx 7.8\,M_\odot$. At other precessional phases, when a stream dominates, $K \approx 80$ km/s and $\phi_0 = 0.12 \pm 0.04$. The half-amplitude of the He II radial velocity based on all data without regard to the precessional phase is ≈ 120 km/s. The gaseous stream that radiates the He II line lags the relativistic object in orbital phase by $\Delta\phi \approx 0.1$, and is probably directed from the star toward the disk. Later, Fabrika et al. (1997a) confirmed using additional observations that the half-amplitude of the orbital variability of the He II line at precessional phases near T_3 is $K = 176 \pm 13$ km/s. In their analysis of eclipses of this line profile based on coordinated observations near primary eclipses, Goranskii et al. (1997) and Fabrika et al. (1997b) were able to distinguish a broad component of the line forming in the disk (it is fully eclipsed at phase $\phi = 0.0$) and a narrow component that forms in the flow (it is partially eclipsed at phase $\phi = 0.1$, and its radial velocity corresponds to the radial-velocity curve of the flow).

D'Odorico et al. (1991) reported a low mass function for SS433, with a half-amplitude for the He II variability of $K \approx 112$ km/s and a phase for upper conjunction of the source based on their radial-velocity curve of $\phi_0 = 0.08$. This half-amplitude for the radial velocities leads to low values for the mass function, $f(M) \propto K^3$. D'Odorico et al. (1991) obtained their spectra over 120 days, and this interval did not include precessional phases corresponding to the maximum turning of the disk to face the observer, $0.9 < \psi < 0.1$. Therefore, their radial-velocity curve corresponds to the gaseous flow rather than the relativistic component, as is also clear from its parameters. The data of D'Odorico et al. (1991) is in full agreement with the results of Fabrika and Bychkova (1990) for precessional phases that correspond to times when the disk is viewed nearly edge-on.

It is interesting that Crampton and Hutchings (1981a) did not divide their observations according to precessional phase, but nevertheless obtained a large amplitude for the He II line shifts. Their spectra were obtained in two observing seasons, both of which included the precessional-phase interval $0.9 < \psi < 0.1$, when the contribution of the accretion disk to the He II line emission is maximum (we do not know the specific selection of spectra used by these authors for their radial-velocity analysis). It is possible that this meant that Crampton and Hutchings (1981a) were able to record the real radial-velocity curve of the accretion disk. They note that the intrinsic half-amplitude of the shifts of this line may be somewhat smaller, but it cannot be less than 150 km/s.

Thus, it has been fairly firmly established (Crampton and Hutchings 1981a; Fabrika and Bychkova 1990; Fabrika et al. 1997a) that the mass function for SS433 derived from the orbital motion of the accretion disk (the He II line) is in the range 7–10 $M_\odot$. Recently, Gies et al. (2002a) discovered that the phase shift in the orbital variability of the radial velocities of the C II λ7231, 7236 emission lines in the spectrum of SS433 corresponds to the orbital position of the accretion disk, with the half-amplitude of the variations being $K \approx 160$ km/s. This confirms the results obtained for the He II line. It is possible that other lines radiating near the relativistic star will be discovered in the near infrared, where the object is fairly bright.

The Component Mass Ratio

The component mass ratio of SS433 $q = M_x/M_o$ has been estimated many times. Based on modeling of the optical eclipses, Antokhina and Cherepashchuk (1987) concluded that the mass ratio was $q \approx 0.25$. However, there is a fairly broad residual minimum allowing larger mass ratios, in particular, $q \sim 0.4$, and possibly even higher values. High values of q were derived by Leibowitz (1984), $\gtrsim 0.8$, and Hirai and Fukue (2001), ≈ 1–1.5. These last authors fit light curves using a thick-accretion-disk model. Similar studies of the shape of X-ray eclipses yield appreciably lower mass ratios. Antokhina et al. (1992) concluded that $q = 0.15$–0.2, with values up to 0.3 possibly being allowed. Kotani et al. (1998) found the ratio $q \approx 0.22$ based on an analysis of eclipses observed by ASCA. If a point source is eclipsed by the star, which fills its critical Roche lobe, the duration of the X-ray eclipses leads to the value $q = 0.15$ (Goranskii et al. 1998b).

X-ray and optical models for the eclipses yield systematically different mass ratios, with an "informal" average of the values being near ≈ 0.3. With this q, the mass function $f(M) = 7$–10 $M_\odot$ yields a mass for the relativistic star of $M_x = 3.5$–5.1 $M_\odot$. However, it is certainly not correct to simply average the various mass-ratio estimates, and we must understand the origins of the differences and devise more complex models for the regions of the optical and X-ray emission in order to explain them. It is also important to learn how to take into account additional absorption in gas flows in the system, which distorts the shape of the X-ray eclipses. Other limitations in this method for determining the mass ratio are imposed by the following two assumptions: (1) the source cannot be larger than the Roche lobe of the compact star (an assumption that we must discard) and (2) the size of the donor corresponds fully to the size of its Roche lobe. A star losing mass at a rate of $\sim 10^{-4} M_\odot/\mathrm{yr}$ could have a very dense and extended atmosphere. Failure to take this into account could lead to underestimation of the mass ratio.

Studies of variations of the orbital period could prove to be an effective method for determining the mass ratio. Goranskii et al. (1998b) found that the period did not vary to within $0\overset{d}{.}00008$ over 17 years of intense observations. Analysis of eclipses using archival material indicates that the period has remained virtually constant for 34 years. This corresponds to an upper limit for the rate of variation of the period of $\dot{P}_{orb} \lesssim 2 \cdot 10^{-7}$. Fabrika et al. (1990) investigated variations of the period as a function of the rate of the mass transfer and mass loss in SS433 under the assumption that all the material up to a rate of $\sim 10^{-4} M_\odot/\mathrm{yr}$ reaches the accretion disk, and that, further, some portion of the gas is lost through the Lagrange point L2 beyond the relativistic component, while the remaining gas is lost from the system in the form of a wind from the inner regions of the disk. They based this analysis on an incorrect determination of the rate of variation of the period (they took large-scale fluctuations in the O–C diagram to be variations in the period). It is now clear that the period of SS433 is surprisingly stable (Goranskii et al. 1998b). Nevertheless, if we suppose that $\dot{P}_{orb} = 0$, the analysis of Fabrika et al. (1990) yields a component-mass ratio of $q = 0.7$–0.8.

It is likely that progress in estimating the masses of the stars in SS433 will come from spectroscopic measurements. If the absorption component in the He I $\lambda 6678$ emission line detected by Gies et al. (2002a) arises in the atmosphere of the donor star, the observed shift of the radial velocity of this component yields the estimate $K_o = 126 \pm 26$ km/s, which, in turn, leads to the mass ratio $q = 0.72 \pm 0.17$ (Gies et al. 2002a). If the mass function is $f(M) \approx 7.8\,M_\odot$, a mass ratio of ≈ 0.7 yields a mass for the relativistic star of $M_x \approx 16\,M_\odot$.

Of course, a number of indirect arguments, one of which is the huge luminosity of SS433, suggest that there is a black hole with a mass of ~ 10–$20\,M_\odot$ in the system. However, without direct measurements of the mass ratio, the compact star must remain only a very likely black-hole candidate. The detection of lines of the secondary component of the system would be an optimal solution to the problem of determining the mass ratio.

Such lines were detected in the recent observations of Gies et al. (2002b), which were conducted during three successive nights that encompassed an eclipse of the accretion disk by the donor star at the precessional phase when the disk is maximally turned toward the observer ($\psi \approx 0$). This last circumstance means that material spreading in the plane of the disk does not screen the donor, and we have the best chance of seeing the photospheric spectrum of this star. Weak absorption lines of Ti II, Fe II, Cr II, Si II, Sr II, Ca I, and Fe I were detected in the blue part of the spectrum (4000–4600 Å), which resembles the spectrum of an evolved star – an A-type supergiant ($T_{eff} \sim 8\,000$ K). These lines became stronger when the disk was maximally eclipsed. The radial velocities of the absorption lines showed an orbital shift that was opposite to the shift shown by the lines emitted by the disk surrounding the compact star. Gies et al. (2002b) present weighty evidence that they have detected the spectrum of the donor star in SS433. Based on the amplitude of the absorption-line shifts and taking into account the mass function derived from the He II line (Fabrika and Bychkova 1990), $f(M) = 7.8\,M_\odot$, Gies et al. (2002b) estimate the mass ratio to be $M_x/M_o = 0.57 \pm 0.11$ and the masses $M_o = 19 \pm 7$ and $M_x = 11 \pm 5$. It is obvious that further more detailed observations in the blue should make it possible to refine the parameters of the components of SS433, but the results of Gies et al. (2002b) have already demonstrated that there is a black hole in this system.

Properties of the Gas Flow from the He II and Hβ Lines

In coordinated spectral and photometric observations of SS433 during eclipses, Goranskii et al. (1997) were able to distinguish three components in the He II λ4686 emission. Components in the He II profile are also clearly visible in the spectra of D'Odorico *et al.* (1991), however continuous observations from night to night near the primary eclipse are required in order to identify these components and trace their variations. In addition, at different precessional phases, the He II components have different appearances and display different behaviour during eclipses (Fabrika *et al.* 1997b). The He II emission consists of a "narrow", essentially Gaussian profile with FWHM ≈ 950 ± 20 km/s and two components forming a broad two-peaked profile.

The narrow He II component is not eclipsed at the center of the photometric eclipses. The region in which this component is emitted experiences partial eclipses (≈ 30–40%) at orbital phase 0.1. The Hβ line consists only of one narrow component (FWHM = 840 ± 40 km/s), which also undergoes partial eclipses with amplitudes of about 15%, but at orbital phases $\phi = 0.1 - 0.25$. The profile and width of the Hβ line and the narrow He II component do not vary during the course of the eclipses. Fabrika *et al.* (1997c) also distinguished eclipses in the Hα line using observations in a narrow filter and in the neighbouring continuum. The eclipses occur at orbital phase $\phi \approx 0.2$, and the depth of the Hα eclipse is about 15%.

The narrow He II component and the hydrogen lines form in a *gaseous flow directed toward the accretion disk*. The half-amplitude of the radial velocities of emission lines emitted in the flow is $K \approx 80$ km/s. The orbital phases at which the regions of line radiation in the flow are at upper conjunction grow from $\phi_0 \approx 0.1$ for He II (the narrow component near precessional phase T_3 or the entire line at other precessional phases) to $\phi_0 = 0.1$–0.2 for the He I lines and $\phi_0 \approx 0.25$ for the hydrogen lines (Crampton *et al.* 1980; Kopylov *et al.* 1989; Goranskii *et al.* 1997; Fabrika 1997; Fabrika *et al.* 1997abc; Gies *et al.* 2002a). We can conclude from this information that the observed flow is very extended. The phase and duration of the He II and Hβ eclipses (Goranskii *et al.* 1997) indicate that the size of the flow is no less than $0.4\,a$ and that, on average, the flow itself is separated from the accretion disk by a distance of $\sim 0.6\,a$, where a is the distance between the components. The flow is directed toward the accretion disk, and the gas temperature in the flow falls with distance from the disk from $(3\text{--}5) \cdot 10^4$ K to $(1\text{--}2) \cdot 10^4$ K. It is possible that the gas is heated by shock waves arising when the flow comes into contact with the accretion disk.

Essentially, all the main emission lines in the spectrum except for the He II lines are radiated in the flow. The profiles of lines emitted in the flow are very strongly distorted by absorption in their blue wings (in the outer wind), and are shifted toward the red by up to $\Delta V_r \sim 200$ km/s, depending on the optical depth of the line and the wind velocity along the line of sight; the magnitude of this shift also strongly depends on the precessional phase (Fabrika 1997; Fabrika et al. 1997a; Gies et al. 2002a). The widths of the flow lines are much larger than the virial velocity of the system. Kopylov et al. (1989) proposed that the flow is optically thick to electron scattering ($\tau \sim 150$), and that the lines are broadened by scattering of the emerging radiation.

The Structure of the Disk and the Central Region from the He II Line

The precession of the disk and its eclipses by the optical star create the rare possibility of directly studying the disk itself and the region where the relativistic jets appear. Here, we must recall that the object surrounding the relativistic star in SS433 is traditionally called a "disk", although it resembles a disk less and less with each new investigation. The wind of the supercritical accretion disk is observed around the relativistic star, and its structure is certainly complex.

The broad, two-peaked He II component is completely eclipsed at the centre of the primary minimum. Based on observations of an eclipse near precessional phase $\psi = 0.95$, Goranskii et al. (1997) established that the blue wing of the broad profile appeared first during the egress of the disk from eclipse ($\phi = 0.1$), with the red wing appearing the following night. In another eclipse at precessional phase $\psi = 0.0$, two peaks with approximately equal intensities appeared simultaneously, when the disk emerged from behind the limb of the star. The distance between the maxima of the two-peaked profile is $\Delta V \approx 1\,500$ km/s. Such a profile cannot belong to lines radiated in the disk. For a Keplerian velocity $\Delta V / 2 \approx 750$ km/s and reasonable masses for the compact star ~ 5 $M_\odot$, the size of the disk in the He II line would be $\sim 10^{11}$ cm. The time for egress from eclipse of such a disk would not be more than two hours, while the observed egress from eclipse of the broad He II component lasts no less than a day.

Goranskii et al. (1997) proposed that the two-peaked He II profile forms in hot, gaseous cocoons surrounding the bases of the approaching and receding jets. The sequence in which the blue and red components appear from behind the limb of the star is consistent with the geometry for the positions of the jets in the standard model in which the precessional and orbital motions are in opposite directions. The velocity of the outflow of gas in these "He II cocoons" was estimated from the known inclination of the jets at the observation epoch to be $V_w(\text{He II}) \approx 1\,500$ km/s. If we indeed see the hot bases of the jets in the He II emission, the distant cocoon radiating the red He II wing is not screened by the accretion disk; i.e., in projection onto the plane of the sky, the distance between the relativistic star and the cocoon is larger than the radius of the disk. Essentially the same situation is observed in the X-ray (see the section "The X-ray Jets") – the body of the disk does not block the receding jet. A full eclipse of both components of the broad He II profile by the star means that the size of the star exceeds the projection of the accretion disk onto the plane of the sky.

Based on a comparison of the times for the emergence of the He II region and the X-ray source in Fe XXV line from behind the limb of the star, Goranskii et al. (1997) found that the size of the He II emission region $(0.25\text{--}0.30\,a)$ exceeds the size of the X-ray emission region $(\approx 0.20\,a)$. The observations of the X-ray eclipse of the Fe XXV line were taken from Ginga data (Kawai et al. 1989). In those observations, the jet X-ray line and the stationary line of weakly ionized iron were not well resolved. It is possible that the He II region surrounds not only the X-ray jets but also the region in which the fluorescent iron line is emitted. It was also noted that the blend of the C III, N III λ_{eff} 4644 lines $(T \sim 30\,000$ K) showed behaviour during eclipses that was similar to that shown by the He II line (narrow line peaks at the centre of eclipses and an overall broadening of the blend during the egress from eclipse).

All these data suggest that the base of the jets can be represented by a cocoon of hot gas enshrouding the region through which the jet passes, with the temperature in the cocoon falling from $\sim 10^8$ K at the axis to $\sim (3\text{--}5) \cdot 10^4$ K at the edges. It is likely that this same region is the source of the polarized UV radiation (Dolan et al. 1997). Observations in several other eclipses (Fabrika et al. 1997b) confirmed that only the narrow He II component remains at the centre of eclipses, with the two-peaked, broad component of the profile appearing during the egress of the disk from eclipse. The pattern for the appearance of the blue and red components of the two-peaked profile varies with the precessional phase, and is not always evident due to missing observations because of variable observing conditions.

The profiles of lines arising in the supercritical disk (the "superdisk") were calculated by Fukue (2000), taking into account the effects of reradiation and screening of the radiation of some parts of the disk by others. The lines are two-peaked, as in an ordinary disk; but if advective motions, in which the velocity of the radial flow toward the centre grows sharply, are taken into account in the dynamics of the disk, the blue component of the profile becomes brighter than the red component. This is a projection effect associated with the fact that part of the hot, inner surface of such a disk is blocked by the edge of the disk. Overall, the orientation of such a superdisk relative to the observer appreciably influences the observed radiation. While the edges of the disk can be essentially dark, the near-polar radiation is strengthened by reradiation effects. The luminosity of the superdisk is $L \sim 9(H/r)L_e$, where H/r is the ratio of the thickness to the radius of the disk, i.e., it depends on the degree of opening of the disk. It is interesting that, for a mass of the compact object $10\,M_\odot$, the luminosity of the superdisk is approximately equal to the observed bolometric luminosity of SS433, $\sim 10^{40}$ erg/s.

It would be tempting to associate the two-peaked He II profile with the superdisk (Fukue 2000), especially since the blue peak is often observed to be brighter than the red peak. However, as is noted above, the size of the region in which this line is radiated appreciable exceeds the Keplerian radius, so that the He II peaks cannot be formed in the disk.

Filippenko et al. (1988) detected Paschen lines ($P_{11} - P_{15}$) with two-peaked profiles in the spectrum of SS433, with the distance between the peaks being $\Delta V \approx 290$ km/s. These observations were carried out at precessional phases that are close to those when the disk is edge-on and outside of eclipses. In addition to the Paschen lines, the Fe II and Hβ lines also had two peaks. The ratio of the intensities in the two peaks changed slightly over three consecutive nights of observations. Filippenko et al. (1988) proposed that the two-peaked lines are manifestations of the accretion disk. The time for such a disk to be eclipsed by the optical star (1–1.5 days) is quite reasonable from an observational point of view, so that it should be possible to test the hypothesis that the double lines arise in the disk by searching for eclipses in the lines at the epochs of primary minima. Variations in the emission-line profiles or in the radial velocities at the level of $\sim$ 100 km/s during eclipses should be relatively easy to detect, but no reports of such effects appear in the literature. The two-peaked ("multi-component") nature of the lines is observed at precessional phases when the disk is edge-on (Crampton and Hutchings 1981b), and it is quite likely that it is associated with the flow of gas from the system through the outer Lagrange point (an excretion disk) in the plane of the accretion disk. Filippenko et al. (1988) put forth this hypothesis as an alternative explanation of the observed two-peaked emission lines. In this case, the total mass of the system should be rather high ($M \gtrsim 40 M_\odot$). This is in agreement with the most recent estimates of the component masses based on spectroscopic data (Gies et al. 2002b).

Precessional Modulation of the Stationary Lines

Crampton and Hutchings (1981ab) noted that the radial velocities of the emission and absorption lines depend on the precessional phase. In their study of the precessional variability of the radial velocities, Fabrika et al. (1997a) concluded that it reduces to a variable absorption component in the blue wing of the emission lines. At precessional phases when the accretion disk is edge-on, the absorption in the blue emission wings increases sharply and shifts toward the centre of the emission line (a P Cygni profile is visible); the remaining line emission turns out to be shifted toward the red. When the disk begins to turn to face the observer (the precessional phase approaches time T_3), the absorption moves away from the emission toward the blue, and the intensity of the absorption decreases, so that the radial velocity of the emission lines is decreased. The radial velocities of some He I lines and of the He II line approach the normal value for the system ($V_r \sim 0$ km/s) as the time T_3 is approached. This is directly related to the effect described above that the real orbital variability of the He II line can be measured only at precessional phases $0.9 < \psi < 0.1$. Accordingly, the intensity of the emission lines grows when the disk faces the observer and decreases when the disk is observed edge-on (Crampton and Hutchings 1981b; Asadullaev and Cherepashchuk 1986; Fabrika et al. 1997a), which is also a consequence of the variable absorption. Thus, at precessional phases when the disk is edge-on, the mean radial velocities of emission lines grow and their intensity decreases, while, on the contrary, the emission-line radial velocities decrease and their intensities grow when the disk partially faces the observer.

The orbital and precessional variability of the radial velocities distort each other. The orbital variability varies appreciably with precessional phase. On the other hand, after correcting for the orbital variability, the scatter of line radial velocities in the precessional curves is appreciably smaller. Certain regularities in the precessional variability of various lines appear, in particular, for the He I, Hβ and He II emission lines: the higher the amplitude of the precessional variability (from 50 to 120 km/s), the lower the mean radial velocity of the lines (from 100 to 200 km/s). These effects also provide evidence that the precessional variability is not associated with real variations of the regions of line emission, but instead with a variable contribution of the absorption in the blue wings of the lines. This is possible if the regions of emission and absorption are spatially separated.

Apart from the precessional and orbital periods in the variations of the emission-line radial velocities, new periods are observed (Fabrika et al. 1997a), the strongest of which is 23.22 days ($K_{23} \approx 115$ km/s). In contrast to the precessional modulation, the variations with these periods are not associated with corresponding variations in the underlying absorption. It may be that this periodicity is a consequence of nodding motions in the accretion flow or spiral shocks in the disk.

Variability of Absorption Lines. Profile of the Velocity of the Disk Wind

The absorption lines in the spectrum of SS433 behave in a surprising fashion. An example of the absorption lines can be seen in

Fig. 1, which shows the weak blue absorption components in emission lines that create P Cygni profiles. The strength of the absorption lines grows sharply when the disk is edge-on; there are two such times during the precessional cycle of SS433, corresponding to the times T_1 ($\psi = 0.34$) and T_2 ($\psi = 0.66$). Accordingly, the absorption lines become stronger twice during each precessional cycle (Crampton and Hutchings 1981b). This should obviously be associated with an increased density of gas lost by the system in the plane of the accretion disk, since the line of sight lies in the plane of the disk at times T_1 and T_2.

The intensity of the absorption lines is also increased at orbital phases $\phi \sim 0.1$, i.e. immediately after eclipses of the accretion disk (Fabrika et al. 1997b; Fabrika 1997). This effect can be seen in Fig. 1. The upper spectrum was obtained nearly at the centre of the primary eclipse, while the lower spectrum was obtained at orbital phase $\phi = 0.096$. The increase in the absorption intensity during the egress from eclipse (when the bright source emerges from behind the limb of the star) is associated with an increase in the gas density along the line of sight in a zone of perturbed wind. The wind from the disk blows at the donor star, and the density of the wind should be enhanced at the interface where there is an interaction between the wind and star and perturbation of the wind. At precessional phases when the disk maximally faces the observer, the absorption lines are generally barely discernible, but they still become substantially stronger during eclipses and immediately afterwards, at orbital phases $\phi = 0.0$–0.2. The higher the wind speed along the line of sight (the closer to precessional phase 0), the earlier the strengthening of the absorption lines begins and ends. The geometry for the perturbations in the flow of gas around the star should indeed depend on the wind velocity. The higher the wind velocity around the star (compared to the constant velocity of the orbital motion), the less should be the curving of the wake in the perturbed wind. Precisely this relation follows from observations. We can see in Fig. 1 that the absorption lines are easily discernible even at the centre of the primary minimum. When these observations were obtained, the wind velocity along the line of sight was $\sim 1\,200$ km/s.

The precession of the accretion disk makes it possible to use the absorption lines to measure the wind velocity in SS433 (Fabrika et al.

1997a) as a function of the polar angle α measured from the disk axis. According to the kinematic model, we are able to study the wind only in the polar-angle interval $60° < \alpha < 90°$. When the disk is edge-on ($\alpha = 90°$), a dense and slow wind is observed ($V_w \sim 100$ km/s), while the wind velocity sharply increases as the angular distance from the plane of the disk increases, reaching values $V_w \sim 1300$ km/s. When the disk is maximally turned to face the observer, the Hβ and He I absorption lines become very weak. Figure 18 shows the velocity of the wind from the accretion disk as a function of the polar angle measured from absorption lines of various elements in the polar-angle interval $60° < \alpha < 90°$. The radial velocities of the absorption lines were measured using data for many precessional cycles. While the hydrogen and He I lines show essentially the same dependence, the iron line (the unblended Fe II λ5169 line) follows this dependence only for velocities ~ 600 km/s, beyond which its radial velocity again begins to decrease, reaching values $V_w \approx 340$ km/s when $\alpha \approx 60°$. The figure also shows the wind velocity indicated by the He II line near angles $\alpha \sim 10°\text{--}20°$ under the assumption that the two-peaked He II profile forms in cocoons enshrouding the bases of the jets. In contrast to the Hβ, He I and Fe II data, the wind velocity indicated by the He II line is not derived from direct measurements.

As the polar angle decreases, the velocity of the gas flowing out from the disk grows sharply from 100–150 km/s for $\alpha = 90°$ to $V_w \gtrsim 1300$ km/s for $\alpha \approx 60°$ (Fabrika et al. 1997a; Fabrika 1997). In this range of α, the wind velocity is well approximated by the relation

$$V_w = (8\,000 \pm 100\,\text{km/s}) \cdot \cos^2\alpha + 150 \pm 10\,\text{km/s}.$$

The available data on the wind are in very good agreement with the picture of gas flowing out from a supercritical accretion disk, for which a model was first described by Shakura and Sunyaev (1973). In this model (see also van den Heuvel 1981; Seifina et al. 1991), the final wind velocity is $V_w \sim (2GM_x/R_{sp})^{1/2}$, where R_{sp} is the spherisation radius of the accretion disk. Adopting $V_w = 1500$ km/s, we find that the spherisation radius in SS433 is $R_{sp} \sim 7 \cdot 10^{10}\,m_6$ cm if the mass of the relativistic star is taken to be $m_6 = M_x/6\,M_\odot$. At distances from the relativistic star exceeding $\sim 7 \cdot 10^{10}\,m_6$ cm, the SS433 accretion disk is "normal", i.e. it may not differ strongly from the accretion disks of cataclysmic variables. Continuing with

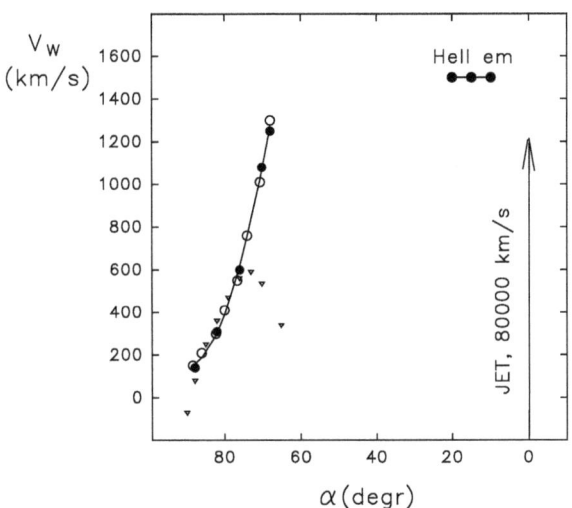

Figure 18. Velocity of the wind outflow from the SS433 accretion disk as a function of the disk polar angle (Fabrika *et al.* 1997a), derived from observations of absorption lines at different orientations of the disk. The open and filled circles on the left show the data for the Hβ and He I λ5015 absorption lines. The triangles show the outflow velocity measured using the Fe II λ5169 absorption line. The reverse behaviour of the velocity derived from the iron line indicates that the fast wind overtakes the slow wind at larger distances from SS433. The terminal mean wind velocity along the line of sight is about $V_w \approx 340$ km/s. The data on the He II emission (the He II cocoon) are model-dependent.

the same simple relationships, the rate at which gas is supplied to the SS433 accretion disk $\dot{M}_a = L_e R_{sp}/GM_x$ turns out to be $\dot{M}_a = 2 L_e/V_w^2 \sim 10^{-3}\, m_6\, M_\odot/\text{yr}$. The critical or Eddington luminosity is $L_e \sim 8 \cdot 10^{38}\, m_6$ erg/s for the same mass of the compact star. The observed bolometric luminosity of SS433 exceeds this critical luminosity by an order of magnitude.

The rate at which gas flows from SS433 is $\dot{M}_e \sim 10^{-4}\, M_\odot/\text{yr}$ (Shklovskii 1981; van den Heuvel 1981). The size of the wind photosphere R_{ph} is determined by the velocity of the outflow, the mass-loss rate and the temperature of the gas (the absorption coefficient). As is noted above, the observed radius of the bright source around the relativistic object in SS433 is $R_{ph} \approx (1.5\text{--}2) \cdot 10^{12}$ cm, and the observed blackbody temperature of the source is $T_{ph} \gtrsim 5 \cdot 10^4$ K. If the outflow from the disk is spherically symmetrical (which is a crude approximation), the mass-loss rate in the wind will be

$$\dot{M}_e \approx 4\pi m_p R_{ph} V_w / \sigma_T \sim 10^{-4}\, M_\odot/\text{yr}.$$

This is close to many other estimates of $\dot{M}_e$ based on independent data, such as IR data (Shklovskii 1981) and radio observations (Blundell et al. 2001). It turns out that the rate at which gas reaches the outer edge of the disk appreciably exceeds the rate at which gas flows from SS433: $\dot{M}_a/(\dot{M}_e + \dot{M}_j) \sim 10$. It is possible that this provides support for models with advective supercritical accretion disks (Eggum et al. 1985, 1988; Okuda 2002), in which an appreciable fraction of the accreted material and radiation is absorbed by the black hole. This would also indirectly confirm that the relativistic star in this system is a black hole.

Structure of Equatorial Outflows in SS433

The absorption lines appear and are sharply enhanced approximately at the times when the disk is viewed edge-on T_1 and T_2 (Crampton and Hutchings 1981b), but in fact, the times when the absorption intensity is maximum are delayed somewhat relative to the precise times T_1 and T_2 by an amount $\Delta\psi(I_{abs}) = 0.15$. Fabrika et al. (1997a) proposed that this delay was associated with the time necessary to accumulate sufficient line-of-sight optical depth in the outflowing gas to form the absorption lines. The emission-line radial-velocity maxima show roughly the same delay. Absorption in the wind distorts the blue side of the emission-line profiles, and the maximum emission-line radial velocities are observed when the line profiles are maximally distorted (at "edge-on" phases). However, the maximum emission-line radial velocities do not occur at phase $\psi = 0.5$ (half way between T_1 and T_2), as might be expected, and show a delay by $\Delta\psi(V_{r,em}) \approx 0.12 \pm 0.07$ averaged over all lines. The minimum intensity due to the precessional variability of the emission lines should be observed when there is maximum absorption in the wind, at phase $\psi = 0.5$, with the intensity maximum at phase $\psi = 0.0$. The delay is also observed here: the Hα intensity minimum lags phase $\psi = 0.5$ by $\Delta\psi(I_{H\alpha}) = 0.19$, according to the data of Asadullaev and Cherepashchuk (1986); the maximum Hα intensity lags phase $\psi = 0.0$ by $\Delta\psi(I_{H\alpha}) \approx 0.1$ (Fabrika et al. 1997c), while the minimum Hβ intensity lags $\psi = 0.5$ by $\Delta\psi(I_{H\beta}) = 0.13$ (Fabrika et al. 1997a). Finally, roughly the same delay is observed for the behaviour of the absorption-line radial velocities with varying precessional phase. When we observe SS433 when the accretion disk is oriented edge-on, the absorption-line radial velocities are maximum (-100 km/s); after the second edge-on position (T_2), the disk turns toward the observer and the absorption radial velocity decreases appreciably. The maximum absorption radial velocity is observed not at time T_2 ($\psi = 0.66$), but later by $\Delta\psi(V_{r,\,abs}) \approx 0.11$ averaged over all the lines. The phase of the maximum radial velocity is somewhat different for each absorption line, which was taken into account when constructing Fig. 18.

All these delays are very similar. It is important that they have been measured for different lines and using different parameters of the lines, but all these delay effects have one origin – variability of the line absorption. The delay is determined by the time required for the accumulation of sufficient optical depth in the gas flowing out in the plane of the accretion disk to form the absorption lines. In the plane of the disk, the gas flows out with the velocity $V_w \approx 100 - 150$ km/s, as follows from the maximum (but negative) Hβ and He I absorption-line radial velocities, but the maximum radial velocity measured from the Fe II absorption lines varies in the range $V_r = +50$ to -150 km/s. The Fe II lines have a very weak emission component, so that their radial velocities can be measured most accurately, without significant systematic errors. On the other hand, the orbital variability could introduce appreciable distortion, and we believe for this reason that the flow velocity in the plane of the accretion disk has been estimated only very approximately, $V_w \sim 100$ km/s.

The behaviour of the radial velocities and intensities of the absorption lines with precessional phase indicates that the outer parts of the SS433 accretion disk participate in the precessional motion. Indeed the outflow of gas from the outer edge of the disk, and therefore the outer edge of the disk itself, participate in the precessional motion. This means that the angular momentum of the material flowing from the donor star also precesses, providing independent support for slaved precession of the SS433 accretion disk and driven precession of the donor star (Shakura 1972; Roberts 1974; van den Heuvel et al. 1980; Whitmire and Matese 1980; Katz 1980; Hut and van den Heuvel 1981). Accordingly, the inner parts of the disk where the rapid wind and jets form also precess.

In the plane of the disk, the system can lose gas most efficiently through the libration point L2, and this loss of gas is associated with the removal of angular momentum during the formation of the disk. Through the L2 point the system can lose at least half of the total gas supplied by the donor star overfilling its Roche lobe (Sawada et al. 1986). An additional source of loss of angular momentum appears above the plane of the disk – the supercritical wind. The gas lost by the system through L2 leaves the system along a winding spiral. It is likely that this is the outflow inferred by Filippenko et al. (1988) from the two-peaked Paschen emission-line profiles. If this represents an excretion disk, the velocity of its rotation (plus expansion) is ~ 150 km/s, in good agreement with absorption-line data. As discussed above, gas flowing out in the plane of the accretion disk is observed in the X-ray (via the absorption of the radiation of the receding jet and the distortion of the orbital light curves), in optical photometric data (distortion of the orbital light curves), and in VLBI images (the central gap and equatorial disk); it has been predicted that this flow could also be detected in the form of an extended Hα disk around SS433 (Fabrika 1993).

The outflow velocity in the plane of the accretion disk in the immediate vicinity of the system is ~ 100 km/s. If we consider the distribution of the wind density along a fixed direction (the line of sight), the outflowing gas should be distributed non-uniformly at small distances from the system r $\lesssim 5 \cdot 10^{13}$ cm, which corresponds to motion with velocities 100–150 km/s over several orbital periods. The regions of enhanced density are modulated with the orbital period, and the distance between them is $(1-1.5) \cdot 10^{13}$ cm. The distance between condensations and the amplitude of the density variations should become smaller with increasing distance from the system, since high-velocity gas ejected from the accretion disk after slower gas (in the same direction) has had time to catch up with this slower gas. At large distances from the system, the wind along the line of sight is also modulated with the precessional period, and the distance between gas condensations is $\approx 5 \cdot 10^{14}$ cm in radius.

The velocity of the wind at distances $\sim 10^{14}$ cm can be estimated from Fig. 18. The radial velocity of the Fe II absorption (like the Hβ and He I absorption) grows as the accretion disk turns toward the observer, but only to –600 km/s ($\alpha \approx 75°$). Higher above the plane of the disk, the wind temperature increases enough that it is probably no longer possible for the Fe II ion to exist. However, further, the wind velocity measured from the Fe II lines begins to decrease, and is only $V_t \approx 340$ km/s at phase $\psi = 0.95$ (47 days after the edge-on epoch T_2). This final wind velocity results from the averaging of the impulses of fast and slow gas moving along the line of sight. When the high-velocity wind catches up to the slow wind emitted earlier in the plane of the accretion disk, the slower wind is compressed. It is at these precessional phases that we observe Fe II absorption at the greatest distances from the source. The mean wind speed V_t is observed at distances $\gtrsim 1.4 \cdot 10^{14}$ cm, covered by the gas moving with this speed over 47 days, in places where the conditions for enhanced Fe II absorption are again created.

In VLBI images of the SS433 jets ("The Radio Jets and W50"), a gap or sharp weakening of the radio emission is observed at the centre (Paragi et al. 1999). The binary system is located on the jet axis, but not exactly at the centre of the gap. The radius of the gap is $\approx 1.8 \cdot 10^{14}$ cm, in projection onto the plane of the sky for a distance to SS433 of 5 kpc. Given that the gas flows out within a rather broad range of angles in the equatorial plane (the angular range due purely to precession of the disk is $\pm 20°$), the material absorbing the radio emission in the central gap is located in the equatorial plane a distance $\sim 3 \cdot 10^{14}$ cm from the source. It is likely that this absorbing material is gas in dense regions of the equatorial wind, observed in optical spectra via the Fe II absorption.

The equatorial VLBI disk (Paragi et al. 1999; Blundell et al. 2001) is observed out to appreciably larger distances from SS433, to $\sim (3\text{--}4) \cdot 10^{15}$ cm. The mechanism for this equatorial radio emission is not entirely clear ("The Radio Jets and W50"); its spectrum is thermal, but the brightness temperature is very high. Additional observations and theoretical studies are required, but it is reasonable to say that the conditions in the extended disk are made suitable for radio emission by the dissipation of energy from shock waves. Due to the precession of SS433, the equatorial wind is modulated by slow (~ 100 km/s) and fast ($\sim 1\,500$ km/s) portions of material. The amplitude of this modulation depends on the angle above the orbital plane. In particular, the slow wind should disappear at angles $\gtrsim \pm 20°$, so that only the fast wind remains at large heights above the orbital plane, and it is this fast wind that compresses the dense equatorial wind. The fragments of equatorial wind ($\sim 1\,200$ km/s) detected by Paragi *et al.* (2002) are consistent with the model of the wind that follows from spectroscopic observations.

Gas Streams in SS433

Figure 19 presents a schematic of the SS433 system. With the exception of the accretion disk itself and the donor star, for which no unambiguous observational manifestations have been detected, all the remaining components of the system have been studied observationally, and are presented roughly to scale. The most recent observations of Gies *et al.* (2002b) indicate that the donor star has an A-type spectrum. It follows from their estimates of the component mass ratio $M_x/M_o = 0.57 \pm 0.11$ that the size of the donor in units of the distance between the components is $R_o = 0.43 \pm 0.02$. If some component of the system is not observed directly (for example, a "hot spot" where a gas stream interacts with the disk), it is not depicted in the schematic. The disk wind is shown by the arrows directly above the disk and behind the optical star. We have not shown the wind photosphere or the flow from the point L2 in order to avoid cluttering the figure. The gas stream to the disk may be drawn too far from the star; it may be somewhat closer, but in that case its size must be proportionally decreased. In the forward part of the stream, which is eclipsed by the star at orbital phase $\phi \approx 0.1$, He II, He I and hydrogen emission lines are formed, while there is only He I and hydrogen line emission further from the disk in the stream (not shown in the figure).

The base of the jet is shown in Fig. 19 as an extended region of X-ray emission surrounded by hot gas radiating in He II lines. The

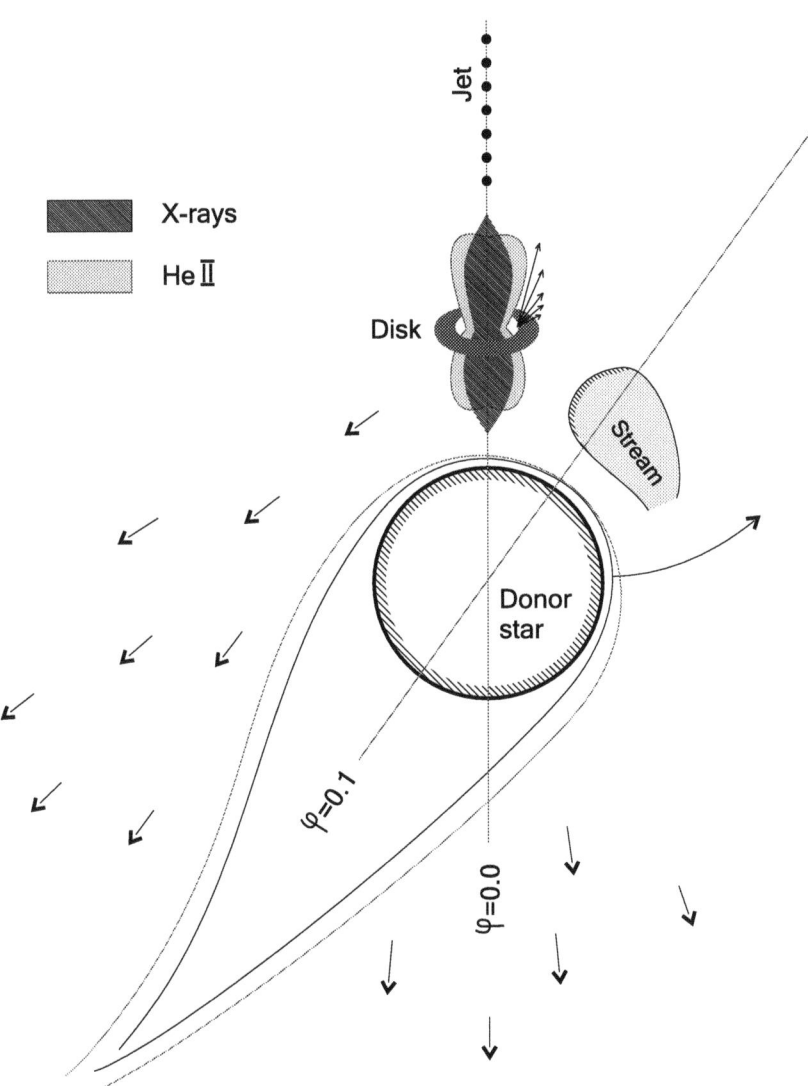

Figure 19. Schematic of the components and gaseous flows in the SS433 system. The "unobserved" accretion disk is shown. The straight arrows denote the wind from the disk, as well as the wind blowing at the donor star.

relative size of this region and of the He II-emission region follow from analyses of eclipse data. At the same time, it is not precisely known what fractions of the eclipsed X-ray flux form in the "slow" wind gas and in the hot jet gas. The model for the cocoon surrounding the bases of the jets will depend on this. The 6.4 keV flourescent line of weakly ionized iron has a small width, FWHM $\lesssim$ 1 000 km/s (Marshall et al. 2002), and could form either within the He II cocoon or in the outer wind. Future observations of X-ray eclipses with high spectral resolution will provide information about the location of the "stationary" X-ray source.

SS433 and Microquasars

Microquasars

In this section, we will briefly describe the properties of microquasars as the closest relatives of SS433, as well as ultra-luminous X-ray sources in other galaxies, which are probably directly related to both SS433 and microquasars. The current literature on microquasars is very extensive; reviews are given by Mirabel and Rodriguez (1999), Greiner (2000), Mirabel (2001), and Fender (2001a, 2002).

Microquasars are believed to be X-ray binaries with relativistic jets. SS433 should probably be considered the prototype for microquasars, since it was the first stellar object in which relativistic jets were discovered. However, the name "microquasar" was first applied to the X-ray binary Sco X-1, which has radio lobes. The radio images of some X-ray binaries with relativistic jets and of radio-loud quasars and radio galaxies are so similar that they are difficult to tell apart without additional information. Therefore, the term "microquasar" was, in turn, chosen to emphasize the morphological similarities of the radio structure of these objects (Mirabel et al. 1992).

As a rule, the term microquasars usually refers to X-ray binaries containing neutron stars or black holes that display jet-like radio activity. This class currently includes slightly fewer than 20 objects, with an additional ten or so candidates (Tsarevsky 2002). For example, the massive X-ray binary Cyg X-1 is also a microquasar. It is believed that jets are always associated with accretion processes, however the variety of the objects considered somewhat smears out the class of microquasars.

Two "classical" microquasars are GRS 1915+105 and GRO J1655–40. Both of these X-ray binaries contain black holes, with masses of $\sim 14\,M\odot$ (Greiner et al. 2001) and $\approx 7.0\,M\odot$ (Orosz and Bailyn 1997), respectively. Classical microquasars are superluminal synchrotron radio sources. The velocities of their jets are 0.92–$0.98\,c$ (Mirabel and Rodriguez 1999). They are transient objects, whose jets are ejected during certain periods of activity; the life times of individual radio blobs in the jets are from several days to several weeks. In active states, the X-ray luminosity is appreciably increased. It is very probable that the jets in these microquasars are leptonic (in contrast to the SS433 jets); i.e., there is acceleration and collimation of relativistic particles in the inner regions near the black hole. Infrared synchrotron radiation is observed during the formation of the relativistic jet (over a time interval of several minutes; Mirabel et al. 1998). Together with the characteristic behaviour of the X-ray emission during flares (see below), this indicates that, in contrast to the situation for SS433, the region in which the jets are generated is fully accessible to observation. It is possible that optical synchrotron radiation during the formation of the jets will also be detected.

The speeds of the jets in microquasars cover a broad range (~ 0.1–$0.9\,c$). However, in nearly all cases, the jet speed cannot be determined accurately, since the distance to the objects and the orientation of the jets are not known. As has already been noted, the observational manifestations of microquasars are extremely varied, and their X-ray variability and quasiperiodic oscillations have been especially well studied. We refer the reader to the reviews listed above, as well as to the recent review of the X-ray continua of microquasars and models for their formation by Poutanen and Zdziarski (2002).

In spite of the large amount of data available, no substantial differences in the jet activity of microquasars containing neutron stars and black holes have been noted. This is striking, since the jet ejection involves the innermost regions, where we would expect the differences between a neutron star and black hole to become very important. However, the ratio of the radio to the X-ray luminosity at the peak of a flare is appreciably higher for black holes than for neutron stars (Fender and Kuulkers 2001). This could be associated both with a higher efficiency for the generation of jets or, for example, with higher absorption of the X-ray emission in the case of black holes.

Fender and Hendry (2000) analysed the radio emission of a number of persistent (non-transient) X-ray binaries containing both black holes (or black-hole candidates) and neutron stars. They concluded that the formation of radio jets requires that the relativistic star not has a strong magnetic field ($< 10^{10}$ G) and that the accretion rate be high ($> 0.1 L_{crit}$); in addition, they found that there is a dramatic physical change in the accretion flow at the time when the jets are launched. In the case of X-ray binaries with neutron stars, the magnetic field must not be strong, so that the accretion flow is not channeled by the field, right to the innermost regions. X-ray pulsars (neutron stars with strong magnetic fields) do not show jet-like radio activity (Fender and Hendry 2000).

A very important circumstance is that the jets in microquasars are ejected during the low/hard X-ray state (Fender 2001a, 2002), which is characterized by a hard, power-law spectrum and strong variability in the X-ray. The presence of radio emission is directly correlated with the source being in the low/hard state. This implies that the intensity of the jet activity can be anticorrelated with the accretion rate.

In the famous flare of the microquasar GRS 1915+105 on September 9, 1997 (Mirabel et al. 1998; Mirabel and Rodriguez 1999), against the background of powerful, short (~ 50 s) oscillations of the X-ray flux, there was first the appreciable dip in the X-ray flux over a time comparable to the oscillation time scale. At the same time, the X-ray spectrum hardened [(13–60 keV)/(2–13 keV)], and the infrared and radio emission also weakened, but more smoothly. During about 7–8 min of the X-ray dip, a sharp, isolated X-ray spike appeared, comprised primarily of soft emission. This is believed to mark the moment when the jet was launched, since the infrared and X-ray fluxes began to grow immediately after the spike (with the oscillations again appearing), and there was a radio flare shortly after the infrared flare. This entire sequence from beginning to end developed over 30–40 min. The X-ray behavior is usually interpreted as reflecting the rapid disappearance (emptying) and subsequent replenishment of the inner accretion disk.

In the massive X-ray binary Cyg X-3, which is believed to probably contain a neutron star in a very close pair with a Wolf–Rayet star (the orbital period is 4.8 h), there is appreciable quenching of the radio emission prior to powerful radio flares (Fender et al. 1997). During strong radio activity, the soft X-ray flux usually grows, interpreted as a substantial growth in the rate at which matter arrives at the accretion disk.

Very interesting behaviour was observed during flares accompanying the ejection of radio jets in Cyg X-3, reminiscent of behaviour shown by GRS 1915+105, but occurring on a substantially longer time scale. The hard X-ray flux at 20–100 keV indicated by BATSE data (McCollough et al. 1999) is anticorrelated with the radio flux in the quiescent state, but becomes clearly correlated with the radio flux during periods of strong activity.

The pattern shown by strong flares of Cyg X-3 (McCollough et al. 1999) is such that there is first a very large weakening of the hard radiation. The radio emission also drops, followed by a powerful radio and X-ray flare or series of several flares, during which the radio and X-ray fluxes are correlated. The ejection of the jet probably occurs at the moment of the sudden dip in the hard X-ray and radio emission.

The existence of such strong correlations suggests (McCollough et al. 1999) that both the accretion disk and the jets contribute to the generation of the hard X-ray radiation. It is also possible that inverse Compton scattering on the radio (synchrotron) electrons plays a determining role in the generation of the hard X-ray radiation.

In the microquasar model developed by Markoff et al. (2001), the synchrotron radiation of the jets and inverse Compton radiation are ascribed a determining role in the formation of virtually the entire spectrum from radio to hard X-ray energies. In any case, it is clear we are dealing with very powerful jets, whose contribution to the total energy released is substantial, no lower than 5% of the total accretion luminosity (Fender 2001b).

Thus, the behaviour of the X-ray emission during flares in microquasars may be determined both by radiation associated with the emerging jet – the synchrotron radiation of the relativistic electrons and inverse Compton scattering of external photons on these electrons, and/or with the emptying of the inner regions of the accretion disk (Greiner 2000). However, it is also possible that the correlation of the X-ray and radio fluxes, the sharp dip in the X-ray emission, and the growth of the hardness ratio of the X-ray emission observed in the X-ray minima of classical microquasars are associated with additional absorption of X-ray emission that arises during the launching of the jets. At this point of the activity, there is a sharp increase in the rate at which gas reaches the inner regions of the accretion disk, or a sudden restructuring of the gaseous flows in these regions. Of course, we must find quantitative and not only qualitative answers to these questions.

The entire period of flare activity (the cycle of activity) in Cyg X-3 occupies 80–100 days, while the characteristic time for the radio flares (and correlated X-ray flares), and also for dips in the radio or X-ray fluxes, is about 10 days.

In classical microquasars (GRS 1915+105), the characteristic times for the development of flares are minutes. Is it likely that we are seeing "naked" relativistic objects in classical microquasars? More precisely, we observe all processes there "in real time". Therefore, observations of microquasars are considered direct testing grounds for the physics of black holes. In the case of Cyg X-3 (and all the more so SS433), this is not possible, since the absorption by the surrounding gas (accretion flows) is appreciably stronger, so that the inner regions are hidden from the observer. Of course, the difference in the time scales cannot be explained purely as an effect of absorption in inner regions of the source. It is likely that the same process are manifest on various time scales: the increase in the accretion rate, restructuring of the accretion disk and flows, and the appearance of jet activity.

Supercritical Transients

Grimm *et al.* (2002) studied the luminosity function of X-ray binaries in our Galaxy using RXTE All Sky Monitor data. The X-ray luminosity depends primarily on the rate of accretion of gas onto the relativistic star, which, in turn, is determined by the rate of loss of gas by the donor. Therefore, we expect a continuous, generally power-law, distribution of X-ray binaries in luminosity, right to the critical luminosity corresponding to the mass of a neutron star ($L_e \sim 2 \cdot 10^{38}$ erg/s) or slightly more than this limiting luminosity. Since the accretion luminosity cannot appreciably exceed the Eddington limit, objects may "accumulate" near this critical luminosity, and we therefore expect a break in the X-ray luminosity function (XLF) at high luminosities. This is approximately what is observed (Grimm et al. 2002; see also references therein to studies of XLFs for other galaxies).

Of course, in the case of "moderate" transient excesses over the critical accretion rate $\dot{M} \leq 10\text{--}100\,\dot{M}_e$, approximately Eddington or slightly super-Eddington sources can appear. Grimm et al. (2002) found that the luminosity function for low-mass X-ray binaries (as earlier, this term concerns only the mass of the donor) has a break near the Eddington luminosity for a neutron star ($\approx 1.4\,M\odot$), and that at least 12 sources showed episodes of supercritical luminosity over the observing time of the RXTE All Sky Monitor program.

The mass-loss rate of the donor and the rate at which mass is captured by the relativistic star depend on many factors. In particular, an increased outflow rate could be the reaction of the donor to accretion activity; however, in general, the donor mass-loss rate is not related to the state of the other component (it is completely unrelated to the presence of an Eddington limit for the luminosity). A sharp increase in the mass transfer could be a consequence of internal processes in the donor atmosphere, the properties of the mass loss, possible precessional motions, the passage of the components through periastron, etc.

In the presence of appreciable transient increases in the mass-accretion rate over short times, an "SS433 syndrome" arises – a sharp dip in the X-ray flux due to absorption in the wind from the accretion disk. Matter is ejected from the system by radiation pressure (Shakura and Sunyaev 1973). In addition, powerful disk-like flows screening the central object or even the entire system can arise. Reprocessing of the X-ray radiation generated in the central regions in the powerful wind should lead to the appearance of a peculiar object that is very weak in the X-ray but bright in the UV and optical. The spectrum of such an object should show broad emission lines that are formed in a wind with a velocity of several thousands of km/s. It is obvious that we would expect the formation of jets, with a sharp enhancement of the radio emission. In addition, it is quite possible that, at such supercritical times, the hard radiation of the object becomes collimated perpendicular to the disk.

The famous giant September 1999 flare of the unusually rapid transient V4641 Sgr (a black hole) was interpreted by Revnivtsev *et al.* (2002ab) as a super-Eddington outburst. The unusually rapid and strong flare of the transient CI Cam (a neutron star or black hole in a pair with a B[e]-supergiant) was also interpreted by Hynes *et al.* (2002) as a supercritical accretion episode. In both cases, the short time of the X-ray flare was associated with the appearance of a wind and absorption of the X-ray radiation. In addition, the corresponding optical flares were unusually bright. The maximum bolometric radiation of such flares should occur in the optical or UV. In both cases, broad optical emission lines were observed, suggesting the formation of a wind in the inner regions of the accretion structure.

Did "heavy and cool" jets like those in SS433 appear during flares of V4641 Sgr and CI Cam? There were no reports of unusual lines during the flares in these objects, but the spectra themselves were very complex and evolved rapidly (see Revnivtsev *et al.* 2002ab and Hynes *et al.* 2002 for references). The data for SS433 suggest that the appearance of cool jets during a supercritical-accretion episode is unlikely, since a persistent channel and a well established disk wind are required to collimate the jets and confine the cool gas clouds in them.

It is even more difficult to answer the question of whether collimated radiation formed during the flares of 4641 Sgr and CI Cam. It is possible that we may have the opportunity of observing a supercritical flare in a "face-on" X-ray transient in the relatively near future. If the solid angle subtended by the opening angle of the channel during the supercritical flare is Ω_c (and, of course, if collimated radiation can emerge in such a situation), then for every $2\pi/\Omega_c$ super-Eddington outbursts, there should be one for which we detect an extremely bright X-ray transient ($L_x \sim 10^{40}$ erg/s).

It may be possible to determine the distances to X-ray transients in our Galaxy if we know the Eddington luminosity for the object (the mass of the relativistic star) and assume that the peak luminosity during a short, supercritical X-ray flare should be very close to the Eddington luminosity. It may also be possible to devise a method for determining the distances to other galaxies based on knowledge of the X-ray luminosity at the break in the XLF, $\sim 2 \cdot 10^{38}$ erg/s (Sarazin et al. 2001). Future X-ray missions should enable the determination of the XLFs of a multitude of galaxies.

A Face-on SS433 and Ultraluminous X-ray Sources in Other Galaxies

The total 2–10 keV X-ray luminosity of the X-ray sources in our Galaxy is $\sim (2-3) \cdot 10^{39}$ erg/s (Grimm et al. 2002), with the total luminosity determined primarily by the few brightest objects. Roughly the same picture is observed for M31 (Makishima et al. 1989), whose total 2–20 keV luminosity is $\sim 5 \cdot 10^{39}$ erg/s.

The brightest X-ray sources in our Galaxy and in the Local Group have X-ray luminosities of a few $\times 10^{38}$ erg/s, while some microquasars reach luminosities of $\sim 3 \cdot 10^{39}$ erg/s at the peaks of flares. However, substantially brighter objects that are not active nuclei (supermassive black holes) are encountered in other galaxies.

One of the most statistically complete surveys of X-ray sources in nearby galaxies was obtained by Roberts and Warwick (2000) using archival ROSAT HRI (High Resolution Imager) data and the list of bright Northern galaxies compiled by Ho et al. (1997). Roberts and Warwick (2000) distinguished 142 non-nuclear X-ray sources in galaxies that were included in HRI observations. Maximum luminosities of their sources reach $L_x \sim 10^{40}$ erg/s. After adding data for M31, Roberts and Warwick (2000) obtained the X-ray luminosity distribution of discrete X-ray sources in 49 spiral galaxies. This distribution normalised to the optical blue luminosity $10^{10} L_\odot$ has the form $dN/dL_{38} = (1.0 \pm 0.2) \cdot L_{38}^{-1.8}$, where L_{38} is the X-ray luminosity in units of 10^{38} erg/s. It also follows from this distribution [Fig. 7 of Roberts and Warwick (2000)] that one source with luminosity $L_x \geq 10^{40}$ erg/s is encountered in a collection of spiral galaxies with total blue luminosity $L_B \geq 10^{12} L_\odot$. This is in good agreement with the fact that there are no sources with this luminosity in the Local Group. The mass of the Local Group is $(1.3 \pm 0.3) \cdot 10^{12} M_\odot$ (Karachentsev et al. 2002).

The most recent studies of the XLFs for sources in other galaxies (Sarazin et al. 2001; Kilgard et al. 2002; Zezas and Fabbiano 2002; Kim and Fabbiano 2003; Colbert et al. 2003; Grimm et al. 2003 and references therein) are based on CHANDRA observations. The spatial resolution and spectral range of CHANDRA enable studies of distributions of point sources, and make it possible to eliminate contamination from supernova remnants to a considerable degree. It was found that these XLFs depend strongly on the starburst activity in the corresponding galaxies. The slopes of the differential XLFs ($dN/dL \propto L^{-\alpha}$) vary from $\alpha \approx 1.5$ in galaxies with strong starburst activity to $\alpha = 2.0$–2.5 in spiral and elliptical galaxies. Grimm et al. (2003), Gilfanov et al. (2003) discuss a universal XLF scaled with the star-formation rate. Its slope is $\alpha \approx 1.6$, and can, in principle, be understood in terms of the population of high-mass X-ray binaries (Postnov 2003). The universal XLF can be fit with a single power law over a very wide luminosity range, $L_x \sim 10^{36}$–10^{40} erg/s, and has a cutoff at $L_x \sim$ a few $\times 10^{40}$ erg/s. The absence of breaks in the universal XLF is strange (Zezas and Fabbiano 2002; Grimm et al. 2003); at least three different populations of objects radiate in this luminosity interval ($\sim 10^{36}$–10^{40} erg/s). Neutron stars cannot be more luminous than $\sim 2 \cdot 10^{38}$ erg/s, and stellar-mass black holes cannot radiate appreciably more than $\sim 10^{39}$ erg/s. In this sense, the universal XLF (Grimm et al. 2003) imposes certain restrictions on the nature of the radiating objects. It may be that we must more accurately consider the anisotropy of the radiation of X-ray binaries and variability of the X-ray sources. Further studies of XLFs in individual galaxies will help reveal more detailed behaviour of the XLFs associated with the star-formation history of the galaxies.

Beginning in 2000, it became clear that ultraluminous X-ray sources ($L_x > (1\text{–}5) \cdot 10^{39}$ erg/s) in galaxies form a separate class of objects. However, can we raise the Eddington luminosity limit in order to avoid considering fundamentally new types of objects?

Grimm et al. (2002) present several reasons why it may be possible to increase this limit somewhat. (i) In the standard theory of Shakura and Sunyaev (1973) with a quasi-flat accretion disk, the radiation flux emerging perpendicular to the plane of the disk exceeds the average value by about a factor of three. (ii) If the chemical composition of the accreting material is poor in hydrogen (the donor is enriched in helium), this will raise the Eddington limit, since the Eddington limit for a helium plasma is twice that of a hydrogen plasma. These two factors can change the classical limit by up to a factor of six. (iii) In theories of supercritical accretion disks, of course, it is possible for the accretion luminosity to exceed the Eddington luminosity. The factor by which the luminosity exceeds the critical value depends on the model (and on whether or not the model is stationary), but even in the earliest supercritical accretion-disk model (Shakura and Sunyaev 1973), the luminosity of the disk was exceeded by the logarithmic factor $\ln(\dot{M}/\dot{M}_{Edd})$, which can be appreciable in appreciably supercritical regimes. In the slim-disk models (Paczynsky and Wiita 1980; Abramowicz et al. 1988), the resulting luminosity is likewise higher than the Eddington luminosity. (iv) There are mechanisms for exceeding the accretion luminosity in the case of accretion onto a neutron star with either a strong or weak magnetic field, due to the formation of specific geometries in the accretion structures (such as a magnetised accretion column).

Nevertheless, the collected observational data on super-Eddington X-ray sources in galaxies (see below) forces us to search for "drastic" solutions to this problem. Either (a) these objects are not super-Eddington, and are black holes with masses of 10^2–$10^4\,M_\odot$, between those for stellar and supermassive black holes – so-called intermediate-mass black holes, or (b) these objects are face-on supercritical accretion disks in binary systems (SS433, microquasars) whose radiation can be collimated by the geometry of a funnel and beamed by the motion of the source along a direction close to the line of sight.

Below, we will consider this second hypothesis – that the radiation of these objects is formed in the funnels of supercritical accretion disks – in more detail. Such objects were first predicted by Katz (1987). We will not be concerned with the relativistic beaming factor, since its magnitude could be determined quite reliably in the case of a face-on SS433 (see the section "Structure and Formation of the Jets"). The beaming factor depends on the spectrum of the emerging radiation, and is ≈ 2. The beaming factors for the relativistic jets of microquasars may be ~ 10–100, depending on the jet velocities.

It was shown in the previous section that the mean radius of the wind photosphere in SS433 is $R_{ph} \sim 1 \div 2 \cdot 10^{12}$ cm, or $(0.2$–$0.5)a$ in terms of the distance between the components. Inside the funnel of the SS433 supercritical disk with its solid angle of Ω_c, the radius (height) of the photosphere is

$$R_{ph,j} \sim \dot{M}_j \sigma_T / \Omega_c m_p V_j \sim 4 \cdot 10^9 \text{cm}$$

for a mass-loss rate in the funnel $\dot{M}_j \sim 5 \cdot 10^{-7} M_\odot/\text{yr}$ and a funnel opening angle $\theta_c \sim 40°$. Collimated radiation could form along such a funnel. For the bolometric luminosity of SS433 $L_{bol} \sim 10^{40}$ erg/s, which is nearly all radiated in the inner regions of the accretion disk, we expect the luminosity of the collimated radiation to also be of the order of $L_c \sim 10^{39}$–10^{40} erg/s.

For an observer able to directly see the central parts of the funnel in SS433, this object would resemble an ultraluminous X-ray source, with a luminosity of $L_x = (2\pi/\Omega_c)L_c \sim 10^{40}$–$10^{42}$ erg/s, up to $\sim 10^4$ times brighter than Cyg X-1. The X-ray flux of an SS433 oriented face-on would vary with a characteristic time scale of $R_{ph,j}/V_j \div R_{ph}/V_j \sim 0.1$–$10^2$ s. The orientation of SS433 relative to the Earth does not enable us to directly study the funnel (although we cannot consider this orientation unfortunate, since we can investigate the binary system itself and the accretion disk by analysing various eclipses in the system). Objects similar to SS433 in other galaxies could be manifest as extremely bright X-ray sources (Katz 1987).

Ultraluminous X-ray sources (ULXs) are indeed observed in other galaxies (Fabbiano 1998; Fabbiano and White 2003). As a rule, they are located in spiral and irregular galaxies, in spiral arms and nuclear regions; i.e., in regions of active star formation. This is consistent with the possibility that ULXs belong to a young stellar population. In our Galaxy, we know the only SS433, and computed evolutionary synthesis models (Lipunov et al. 1996) also predict the presence of only a few objects of this type in a spiral galaxy such as our own. However, in young star-forming regions, the density of the most massive stars (from which a system such as SS433 could form) is enhanced by a factor of a hundred compared to the mean density in the galaxy. Fabrika and Mescheryakov (2001), King et al. (2001) and Koerding et al. (2001) have suggested that ULXs are objects such as SS433 or microquasars oriented face-on. Recently, ULXs have been studied very actively using space-borne instruments. Their main properties (luminosity, spectrum, variability) are consistent with the hypothesis that ULXs are supercritical accretion disks oriented so that our line of sight is close to the disk axis.

Virtually all well studied ULXs display appreciable variability of their X-ray fluxes. This is a strong argument supporting the idea that these objects are supercritical accretion disks oriented face-on. However, when we imagine even such a well studied star as SS433 oriented face-on, it becomes a hypothetical object whose properties (such as its spectrum) can be predicted only with considerable uncertainty.

The frequency with which ULXs are encountered in other galaxies (Roberts and Warwick 2000; Fabrika and Mescheryakov 2001) is reasonably close to the expected value if they are face-on supercritical accretion disks. Fabrika and Mescheryakov (2001) carried out a cross-correlation of sources from the ROSAT All Sky Survey Bright and Faint Source Catalogs (Voges et al. 1999, 2000) and the RC3 galaxy catalog (de Vaucouleurs et al. 1991), which contains 16 741 bright spiral and irregular galaxies. They distinguished 142 sources that were not known to be active nuclei in these galaxies, with 80 of these being located in non-nuclear regions. The X-ray luminosities of these objects are $L_x \sim 10^{39} - 3 \cdot 10^{41}$ erg/s. The estimated frequency with which non-nuclear sources are encountered was ~ 0.05, or one object roughly per every 20 galaxies. This frequency can be understood quantitatively if there are ~ 1 SS433-like objects in each galaxy, with the opening angle of the cone collimating the radiation being equal to $\theta_c = 30° - 40°$ and the orientation of these objects being random. The X-ray spectra (hardness index) of the selected objects show that they are, on average, hard sources.

The source samples of Roberts and Warwick (2000) and Fabrika and Mescheryakov (2001) differ significantly. The former sample was constructed using pointed HRI observations of a relatively small number of bright galaxies ($B < 12.5$), making it possible to study sources with luminosities $\sim 10^{38}$–10^{39} erg/s in detail. However, observations of a larger number of galaxies are required to collect an appreciable number of substantially brighter objects, $L_x \geq 10^{40}$ erg/s. Fabrika and Mescheryakov (2001) used catalogs based on the All Sky Survey, and the sample of galaxies ($V < 15.0$) was significantly more representative. However, sources weaker than $\approx 10^{40}$ erg/s could be completely selected only in nearby galaxies (< 11 Mpc). Therefore, this second sample included, on average, the brightest ULXs candidates.

ULXs have also been identified in elliptical galaxies (Colbert and Prak 2002; Colbert et al. 2003). This does not contradict the interpretation of ULXs as microquasars, since the population of microquasars includes many low-mass X-ray binaries, which are present in elliptical galaxies. However the brightest ULXs are detected in interacting and starbursting galaxies, for example the "hyper-ULXs" ($L_x > 10^{41}$ erg/s) in the Cartwheel galaxy (Gao et al. 2003). Nevertheless, it is possible that ULXs do not represent a uniform class of objects.

The spectra of ULXs are very similar to those of X-ray binaries, and are sometimes well fit by a so-called multicolor disk blackbody model (kT $\sim$ 1–3 keV), although a more complex description of the spectrum is often required (Okada et al. 1998; Makishima et al. 2000; Kotoku et al. 2000; Kubota et al. 2002; Ebisawa et al. 2003). In recent studies a cool multicolor disk model (kT $\sim$ 0.1–0.3 keV) plus a power–law component are required (Miller et al. 2003abc) for the best representation of the ULXs spectra. Some ULXs show very soft or extremely steep X-ray spectra (Fabbiano et al. 2003a; Cagnoni et al. 2003). Like X-ray binaries, ULXs can display transitions between the soft/high and hard/low states of the spectrum (Kubota et al. 2001; La Parola et al. 2001). As has already been noted, the variability of the X-ray flux is very substantial (Mizuno et al. 2001; Mukai et al. 2003; Fabbiano et al. 2003b; Roberts and Colbert 2003), and can reach a factor of two over a time of about one hour. It is likely that studies of variability on shorter time scales are limited only by the sensitivity of modern detectors. There have even been reports of periodic variability on time scales of hours to days in some ULXs (Bauer et al. 2001; Sugiho et al. 2001; Liu et al. 2002a) and short ($\sim$ 20 sec) quasi–periodical oscillations (Strohmayer and Mushotzky 2003).

Radio emission has been detected from a ULX in the galaxy NGC 5408 (Kaaret et al. 2003), with reports that the X-ray, radio and optical fluxes of this object are consistent with the expectations for beamed emission from a relativistic jet. It is very important to obtain optical identifications for ULXs, since they could help provide more certain answers about the nature of these sources. When ULXs are identified, it is with very weak objects with optical magnitudes of 20–25, often located in nebulae (Miller 1995; Roberts et al. 2001; Liu et al. 2002b; Wang 2002; Wu et al. 2002; Roberts et al. 2003; Holt et al. 2003; Zampieri et al. 2003). They are usually blue objects, and some are young clusters (Goad et al. 2002; Zezas et al. 2002).

Bubble–like nebulae around ULXs are frequently detected (Pakull and Mirioni 2003; Roberts et al. 2003). In the region of the ULX in the galaxy Holmberg II, Pakull and Mirioni (2001) detected a nebula radiating in the He II λ4886 excited by the X-ray source. They concluded that there was no strong beaming along the line of sight. However, strong collimation of the radiation or beaming is not required to understand these objects. Thus, the available radio and optical identifications support the hypothesis that ULXs are objects with supercritical accretion disks or microquasars, or at least do not contradict this hypothesis.

An alternative model for ULXs is that they contain intermediate-mass black holes (IMBHs) with masses $\sim 10^3 \, M_\odot$ (Colbert and Mushotzky 1999; van der Marel 2003; Miller and Colbert 2003), which could have formed from the very first (Population III) generation of stars (Madau and Rees 2001) or in globular clusters (Miller and Hamilton 2002). Such black holes could accrete interstellar gas and become bright X-ray sources with luminosities $\sim 10^{40}$ erg/s if the surrounding gas is sufficiently dense ($n > 10^2$–10^3 cm^{-3}) and the velocity of the IMBHs relative to this gas is sufficiently low ($\Delta V < 10$ km/s). These last two conditions substantially limit the number of IMBHs that are accessible to observation.

What criteria can be proposed to distinguish between these two alternative models for ULXs? Investigations of the nebulosities surrounding these sources should include searches for evidence of dynamical interactions between jets that may be emerging from the objects and the interstellar gas. By analogy with SS433 (see the section "The Radio Jets and W50"), we may expect perturbations of the interstellar medium with amplitudes of tens of km/s on scales of tens of parsec. Such features around ULXs could easily be detected in galaxies at distances of up to ~ 10 Mpc, even in ground-based observations. On the other hand, IMBHs can only ionize the interstellar medium, not dynamically perturb it. The radius of the region in which interstellar gas would be captured around a black hole with a mass of $10^3 M_\odot$ moving with a relative speed of $\Delta V = 10$ km/s is only 0.1 pc.

As is mentioned above, one possible test is to search for variability. Black holes are not able to produce strong variability on time scales appreciably shorter than a few $\times$ 0.01 s $(M_{BH}/M_\odot)$ (Sunyaev and Revnivtsev 2000). If we suppose that the IMBHs in ULXs radiate less than the Eddington limit, it is very unlikely that there will be brightness variability associated with these black holes on time scales < 1 s L_{40}, where the X-ray luminosity is expressed in units of 10^{40} erg/s. Detailed studies of the variability of ULXs on such short time scales must await the next generation of X-ray telescopes.

A critical experiment that could enable identification of ULXs with SS433-like objects oriented face-on would be observations indicating the existence of a funnel in the supercritical accretion disk. In this case, the presence of very broad X-ray absorption lines with complex profiles is predicted. These absorption bands should belong to hydrogen- and helium-like ions of the most abundant heavy elements (Fe, S, Si, Mg and others), and should extend from the Kc to the Kα energies of the corresponding ions and transitions. Thanks to the Doppler decrease in the optical depth of the material accelerated in the funnel, it may be possible to study the funnel down to the depth of the photosphere via observations of these absorption lines.

Variations of the gas parameters along the funnel – its velocity, density, temperature and volume filling factor – could make the absorption-line profiles appreciably more complex, necessitating the use of X-ray spectra with high signal/noise ratios in searches for these lines. For example, if, as is proposed in the section "The Structure and Formation of the Jets", the gas in the inner parts of the funnel is first accelerated to velocities $\sim 10^{10}$ cm/s and then decelerates due to the action of the wind from the walls, such that its velocity acquires the value $0.26c$, we expect the presence of a very broad Kα absorption line shifted toward higher energies, with the blue wing of this line extending to energies corresponding to the Kc threshold. Such shallow, broad absorption lines could distort the continuum near the Kα–Kc energies.

In essence, the predicted complexity of the dependence of the absorption-line profiles on the structure of the funnel and mecha-

THE JETS IN SS433 137

nisms for acceleration and collimation of the gas in the funnel present excellent opportunities for direct probing of these structures in supercritical accretion disks and for studies of mechanisms for formation of the jets.

Acknowledgements

The author thanks T.R. Irsmambetova, A.A. Panferov and K.A.Postnov for discussions, and O.N. Sholukhova, N.S. Fabrika, A.E. Surkov, R.A. Karimova and E.A. Barsukova for help with the preparation of the manuscript. The author is especially grateful to T. Kotani, Z. Paragi, V. P. Goranskii and H.L. Marshall for presenting figures, to Z. Paragi for useful comments concerning the radio data, to G. Tsarevsky for providing a list of microquasars, and to R. Sunyaev for very valuable comments. This work was supported by the Russian Foundation for Basic Research (projects N 03-02-16341) and a grant from the State Russian Program "Astronomy".

REFERENCES

1. Abell, G.O. and Margon, B. 1979, *Nature* 279, 701.
2. Abramowicz, M., Czerny, B., Lasota, J. and Szuszkiewicz, E. 1988, *Astrophys. J.* 332, 646.
3. Abramowicz, M.A., Igumenshchev, I.V., Quataert, E. and Narayan, R. 2002, *Astrophys. J.* 565, 1101.
4. Anderson, S.F., Grandi, S.A. and Margon, B. 1983, *Astrophys. J.* 273, 697.
5. Antokhina, E.A. and Cherepashchuk, A.M. 1987, *Sov. Astron.* 31, 295.
6. Antokhina, E.A., Seifina, E.V. and Cherepashchuk, A.M. 1992, *Sov. Astron.* 36, 143.
7. Arav, N. and Belelman, M. 1992, *Astrophys. J.* 401, 125.
8. Arav, N. and Belelman, M. 1993, *Astrophys. J.* 413, 700.
9. Asadullaev, S.S. and Cherepashchuk, A.M. 1986, *Sov. Astron.* 30, 57.
10. Aslanov, A.A., Cherepashchuk, A.M., Goranskij, V.P., Rakhimov, V.Yu. and Vermeulen, R.C. 1993, *Astron. Astrophys.* 270, 200.
11. Band, D.L. 1987, *Publ. Astr. Soc. Pac.* 99, 1269.
12. Band, D.L. and Grindlay, J.E. 1984, *Astrophys. J.* 285, 702.
13. Band, D.L. and Grindlay, J.E. 1986, *Astrophys. J.* 311, 595.

14. Bauer, F.E., Brandt, W.N., Sambruna, R.M., Chartas, G., Garmire, G.P., Kaspi, S. and Netzer, H. 2001, *Astron. J.* 122, 182.
15. Baykal, A., Anderson, S.F. and Margon, B. 1993, *Astron. J.* 106, 2359.
16. Begelman, M.C., Hatchett, S.P., McKee, C.F., Sarazin, C.L. and Arons, J. 1980, *Astrophys. J.* 238, 722.
17. Begelman, M. and Rees, M.J. 1984, *Mon. Not. R. Astron. Soc.* 206, 209.
18. Begelman, M.C., Blandford, R.D. and Rees, M.J. 1984, *Rev. Mod. Phys.* 56, N 2, 1.
19. Bisikalo, D.V., Boyarchuk, A.A., Kuznetsov, O.A. and Chechetkin, V.M. 1999, *Astron. Rep.* 43, 587.
20. Blandford, R.D. and Königl, A. 1979, *Astrophys. J.* 232, 34.
21. Blundell, K.M., Mioduszewski, A.J., Podsiadlowski, P., Muxlow, T.W.B. and Rupen, M.P. 2001, *Astrophys. J.* 562, L79.
22. Blundell, K.M., Rupen, M.P., Mioduszewski, A.J., Muxlow, T.W.B. and Podsiadlowski, P. 2002, In *'New Views on Microquasars"*, *the Fourth Microquasars Workshop*, Ph. Durouchoux, Y. Fuchs, J. Rodriguez (eds). *Center for Space Physics, Kolkata (India), p. 249;* astro-ph/0209365.
23. Bodo, G., Ferrari, A., Massaglia, S. and Tsinganos, K. 1985, *Astron. Astrophys.* 149, 246.
24. Bodo, G., Ferrari, A., Massaglia, S. and Brinkmann, W. 1988, *Astrophys. Lett. Commun.* 27, 5.
25. Bohannan, B. and Crowther, P.A. 1999, *Astrophys. J.* 511, 374.
26. Bonsignori-Facondi, S.R., Padrielli, L., Montebugnoli, S. and Barbieri, R. 1986, *Astron. Astrophys.* 166, 157.
27. Borisov, N.V. and Fabrika, S.N. 1987, *Sov. Astron. Lett.* 13, 200.
28. Brinkmann, W., Fink, H.H, Massaglia, S., Bodo, G. and Ferrari, A. 1988, *Astron. Astrophys.* 196, 313.
29. Brinkmann, W., Kawai, N. and Matsuoka, M. 1989, *Astron. Astrophys.* 218, L13.
30. Brinkmann, W., Kawai, N., Matsuoka, M. and Fink, H.H. 1991, *Astron. Astrophys.* 241, 112.
31. Brinkmann, W., Aschenbach, B. and Kawai, N. 1996, *Astron. Astrophys.* 312, 306.
32. Brinkmann, W. and Kawai, N. 2000, *Astron. Astrophys.* 363, 640.
33. Brown, J.C., Cassinelli, J.P. and Collins, G.W. II. 1991, *Astrophys. J.* 378, 307.
34. Brown, J.C. and Fletcher, L. 1992, *Astron. Astrophys.* 259, L43.
35. Brown, J.C., Mundell, C.G., Petkaki, P. and Jenkins, G. 1995, *Astron. Astrophys.* 296, L45.
36. Bursov, N.N. and Trushkin, S.A. 1995, *Astron. Lett.* 21, 145.
37. Cagnoni, I., Turolla, R., Treves, A., Huang, J.-S., Kim, D. W., Elvis, M. and Celotti, A. 2003, *Astrophys. J.* 582, 654.

38. Calvani, M. and Nobili, L. 1981, *Astrophys. Space Sci.* 79, 387.
39. Cantó, J., Tenorio-Tagle, G. and Różyczka, M. 1988, *Astron. Astrophys.* 192, 287.
40. Ciatti, F., Mammano, A. and Vittone, A. 1978, *IAU Circ.* N 3305, 3.
41. Ciatti, F., Mammano, A. and Vittone, A. 1981, *Astron. Astrophys.* 94, 251.
42. Chakrabarti, S.K. and Matsuda, T. 1992, *Astrophys. J.* 390, 639.
43. Chakrabarti, S.K. Goldoni, P., Wiita, P.J., Nandi, A. and Das, S. 2002, *Astrophys. J.* 576, L45.
44. Chattopadhyay, I. and Chakrabarti, S.K. 2002, *Mon. Not. R. Astron. Soc.* 333, 454.
45. Cherepashchuk, A.M. 1981, *Mon. Not. R. Astron. Soc.* 194, 761.
46. Cherepashchuk, A.M., Aslanov, A.A. and Kornilov, V.G. 1982, *Sov. Astron.* 26, 697.
47. Cherepashchuk, A.M. 1989, *Astrophys. Space Phys. Rev.* 7, 185.
48. Cherepashchuk, A.M., Bychkov, K.V. and Seifina, E.V. 1995, *Astrophys. Space Sci.* 229, 33.
49. Cherepaschuk, A. 2002, *Space Sci. Rev.* 102, 23.
50. Cherepashchuk, A.M., Sunyaev, R.A., Seifina, E.V., Panchenko, I.E., Molkov, S.V. and Postnov, K.A. 2003, astro-ph/0309140.
51. Clark, D.H. and Murdin, P. 1978, *Nature* 276, 45.
52. Clark, D.H. 1985, *The Quest for SS433*. Viking, New York.
53. Colbert, E.J.M. and Mushotzky, R.F. 1999, *Astrophys. J.* 519, 89.
54. Colbert, E.J.M. and Ptak, A.F. 2002, *Astrophys. J. Suppl. Ser.* 143, 25.
55. Colbert, E., Heckman, N., Ptak, A., Strickland D. and Weaver, K. 2003, astro-ph/0305476.
56. Collins, G.W., II 1985. *Mon. Not. R. Astron. Soc.* 213, 279.
57. Collins, G.W., II and Newsom, G.H. 1986, *Astrophys. J.* 308, 144.
58. Collins, G.W., II and Newsom, G.H. 1988, *Astrophys. J.* 331, 486.
59. Collins, G.W. II and Scher, R.W. 2002, *Mon. Not. Roy. Astron. Soc.* 336, 1011.
60. Crampton, D., Cowley, A.P. and Hutchings, J.B. 1980, *Astrophys. J.* 235, L131.
61. Crampton, D. and Hutchings, J.B. 1981a, *Astrophys. J.* 251, 604.
62. Crampton, D. and Hutchings, J.B. 1981b, *Vistas Astron.* 25, 13.
63. Crowther, P.A. and Smith, L.J. 1997, *Astron. Astrophys.* 320, 500.
64. D'Odorico, S., Oosterloo, T., Zwitter, T. and Calvani, M. 1991, *Nature* 353, 329.
65. Davidson, K. and McCray, R. 1980, *Astrophys. J.* 241, 1082.
66. de Vaucouleurs, G., de Vaucouleurs, A., Corwin, H.G., Jr., Buta, R.J., Paturel G., Fouque P. 1991, *Third Reference Cataloque of Bright Galaxies*. Springer-Verlag, Berlin, Heidelberg, New York.
67. Dolan, J.F., Boyd, P.T., Fabrika, S., Tapia, S., Bychkov, V.,

Panferov, A.A., Nelson, M.J., Percival, J.W., van Citters, G.W., Taylor, D.C. and Taylor, M.J. 1997, *Astron. Astrophys.* 327, 648.
68. Dopita, M.A. and Cherepashchuk, A.M. 1981, *Vistas Astron.* 25, 51.
69. Drake, S.A. and Ulrich, R.K. 1980, *Astrophys. J. Suppl. Ser.* 42, 351.
70. Dubner, G.M., Holdaway, M., Goss, W.M. and Mirabel, I.F. 1998, *Astron. J.* 116, 1842.
71. Ebisawa, K., Życki, P., Kubota, A., Mizuno, T. and Watarai, K. 2003, astro-ph/0307392.
72. Efimov, Yu.S., Shakhovskoi, N.M. and Piirola, V. 1984, *Astron. Astrophys.* 138, 62.
73. Eggum, G.E., Coroniti, F.V. and Katz, J.I. 1985, *Astrophys. J.* 298, L41.
74. Eggum, G.E., Coroniti, F.V. and Katz, J.I. 1988, *Astrophys. J.* 330, 142.
75. Eikenberry, S.S., Cameron, P.B., Fierce, B.W., Kull, D.M., Dror, D.H., Houck, J.R. and Margon, B. 2001, *Astrophys. J.* 561, 1027.
76. Fabbiano, G. 1998, In *"Hot Universe"*. Proceedings of IAU Symp. N 188, K. Koyama, S. Kitamoto, M. Itoh (eds). Kluwer Acad. Press, Dordrecht, p.93.
77. Fabbiano, G. and White, N.E. 2003, astro-ph/0307077.
78. Fabbiano, G., King, A.R., Zezas, A., Ponman, T.J., Rots, A. and Schweizer, F. 2003a, *Astrophys. J.* 591, 843.
79. Fabbiano, G., Zezas, A., King, A.R., Ponman, T.J., Rots, A. and Schweizer, F. 2003b, *Astrophys. J.* 584, L5.
80. Fabian, A.C. and Rees, M.J. 1979, *Mon. Not. R. Astron. Soc.* 187, 13P.
81. Fabrika, S.N. 1984, *Sov. Astron. Lett.* 10, 16.
82. Fabrika, S.N. and Borisov, N.V. 1987, *Sov. Astron. Lett.* 13, 279.
83. Fabrika, S.N., Kopylov, I.M. and Shkhagosheva, Z.U. 1990, preprint N 61 of Special Astrophysical Observatory.
84. Fabrika, S.N. and Bychkova, L.V. 1990, *Astron. Astrophys.* 240, L5.
85. Fabrika, S.N. 1993, *Mon. Not. R. Astron. Soc.* 261, 241.
86. Fabrika, S.N. 1997, *Astrophys. Space Sci.* 252, 439.
87. Fabrika, S.N., Bychkova, L.V. and Panferov, A.A. 1997a, *Bull. Spec. Astrophys. Obs.* 43, 75.
88. Fabrika, S.N., Goranskij, V.P., Rakhimov, V.Y., Panferov, A.A. Bychkova, L.V., Irsmambetova, T.R., Shugarov, S.Y. and Borisov, G.V. 1997b, *Bull. Spec. Astrophys. Obs.* 43, 109.
89. Fabrika, S.N., Panferov, A.A., Bychkova, L.V. and Rakhimov, V.Yu. 1997c, *Bull. Spec. Astrophys. Obs.* 43, 95.
90. Fabrika, S.N. 1998, Doct. Diss. Special Astrophysical Observatory, RAS.

91. Fabrika, S. and Mescheryakov, A. 2001, In *"Galaxies and their Constituents at the Highest Angular Resolution"*. IAU Symp. N 205, R.T. Schilizzi (ed.), Manchester, United Kingdom, p.268; astro-ph/0103070.
92. Fabrika, S.N. and Irsmambetova T.R. 2002; In *"New Views on Microquasars"*, the Fourth Microquasars Workshop, Ph. Durouchoux, Y. Fuchs, J. Rodriguez (eds). Center for Space Physics, Kolkata (India), p.268; astro-ph/0207254.
93. Falomo, R., Boksenberg, A., Tanzi, E.G., Tarenghi, M. and Treves, A. 1987, *Mon. Not. R. Astron. Soc.* 224, 323.
94. Fejes, I., Schilizzi, R.T. and Vermeulen, R.C. 1988, *Astron. Astrophys.* 189, 124.
95. Feldman, P.A., Purton, C.R., Stiff, T. and Kwok, S. 1978, *IAU Circ.* N 3258, 1.
96. Fender, R. 2001a, in *"High Energy Gamma-Ray Astronomy"*, F.A. Aharonian and H.J. Völk (eds). American Institute of Physics Proc., 558, p.221; astro-ph/0101233.
97. Fender, R.P. 2001b, *Mon. Not. R. Astron. Soc.* 322, 31
98. Fender, R. 2002, in *"Relativistic Flows in Astrophysics"*, A.W. Guthmann, M. Georganopoulos, A. Marcowith and K. Manolakou (eds); *Lecture Notes in Physics* 589, 101.
99. Fender, R.P., Hendry, M.A. 2000, *Mon. Not. R. Astron. Soc.* 317, 1.
100. Fender, R.P., Kuulkers, E. 2001, *Mon. Not. R. Astron. Soc.* 324, 923.
101. Fender, R.P., Bell Burnell, S.J., Waltman, E.B., Pooley G.G., Ghiggo, F.D. and Foster, R.S. 1997, *Mon. Not. R. Astron. Soc.* 288, 849.
102. Fender, R., Rayner, D., Norris, R., Sault, R.J. and Pooley, G. 2000, *Astrophys. J.* 530, L29.
103. Ferrari, A., Trussoni, E., Rosner, R. and Tsinganos, K. 1985, *Astrophys. J.* 294, 397.
104. Fiedler, R.L., Johnston, K.J., Spencer, J.H., Waltman, E.B., Florkowski, S.R., Matsakis, D.N., Josties, F.J., Angerhofer, P.E., Klepczynski, W.J. and McCarthy, D.D. 1987, *Astron. J.* 94, 1244.
105. Filippenko, A.V., Romani, R.W., Sargent, W.L.W. and Blandford, R.D. 1988, *Astron. J.* 96, 242.
106. Frasca, S., Ciatti, F. and Mammano, A. 1984 *Astrophys. Space Sci.* 99, 329.
107. Fuchs, Y., 2002, astro-ph/0207429.
108. Fuchs, Y., Koch-Miramond, L. and Ábrahám, P. 2002, in *"Neutron Stars in Supernova Remnants"*, P.O. Slane and B.M. Gaensler (eds), ASP Conf. Ser. N 271. ASP, San Francisco, p.369; astro-ph/0112339; in *Proceedings of the 4th Microquasar Workshop*, Ph. Durouchoux, Y. Fuchs and J. Rodrigueez (eds).

Center for Space Physics, Kolkata, p.269; astro-ph/0208432.
109. Fukue, J. 1987a, *Publ. Astron. Soc. Japan* 39, 679.
110. Fukue, J. 1987b, *Publ. Astron. Soc. Japan* 39, 895.
111. Fukue, J., Nakashima, R., Arimoto, J., Awano, Y., Honda, S., Ishikawa, K., Kato, T., Kawai, N., Matsumoto, K., Okugami, M., Sakaguchi, T., Tajima, Y., Tanabe, K., Tsuda, K., Watanabe, Y., Yamada, Y. and Yokoo, T. 1997, *Publ. Astron. Soc. Japan* 49, 93.
112. Fukue, J. 2000, *Publ. Astron. Soc. Japan* 52, 829.
113. Gao, Yu., Wang, Q.D., Appleton, P.N. and Lucas, R.A. 2003, *Astrophys. J.* 596, L171.
114. Geldzahler, B.J., Share, G.H., Kinzer, R.L., Magura, J., Chupp, E.L. and Rieger, E. 1989, *Astrophys. J.* 342, 1123.
115. Gies, D.R., McSwain, M.V., Riddle, R.L., Wang, Z., Wiita, P.J. and Wingert, D.W. 2002a, *Astrophys. J.* 566, 1069.
116. Gies, D.R., Huang, W. and McSwain, M.V. 2002b, *Astrophys. J.* 578, L67.
117. Giles, A.B., King, A.R., Jameson, R.F., Sherrington, M.R., Hough, J.H., Bailey, J.A. and Cunningham, E.C., 1980, *Nature* 286, 689.
118. Gilfanov, M., Grimm, H.-J. and Sunyaev, R. 2003, astro-ph/0309725.
119. Gladyshev, S.A., Goranskii, V.P. and Cherepashchuk, A.M. 1987, *Sov. Astron.* 31, 541.
120. Goad, M.R., Roberts, T.P., Knigge, C. and Lira, P. 2002, *Mon. Not. R. Astron. Soc.* 335, L67.
121. Goranskij, V.P., Kopylov, I.M., Rakhimov, V.Yu., Borisov, N.V., Bychkova, L.V., Fabrika, S.N. and Chernova, G.P. 1987, *Commun. Spec. Astrophys. Obs.* 52, 5.
122. Goranskii, V.P., Fabrika, S.N., Rakhimov, V.Yu., Panferov, A.A., Belov, A.N. and Bychkova, L.V. 1997, *Astron. Rep.* 41, 656.
123. Goranskii, V.P., Esipov, V.F. and Cherepashchuk, A.M. 1998a, *Astron. Rep.* 42, 336.
124. Goranskii, V.P., Esipov, V.F. and Cherepashchuk, A.M. 1998b, *Astron. Rep.* 42, 209.
125. Goranskii, V.P. 2002, private communication.
126. Grandi, S.A. and Stone, R.P.S. 1982, *Publ. Astr. Soc. Pac.* 94, 80.
127. Greiner, J. 2000, in *"Cosmic Explosions: Tenth Astrophysics Conference"*, S.S. Holt and W.W. Zhang (eds). American Institute of Physics Proc., 522, p.307.
128. Greiner, J., Cuby, J.G., McCaughrean, M.J. 2001, *Nature* 414, 522.
129. Grimm, H.-J., Gilfanov, M. and Sunyaev, R. 2002, *Astron. Astrophys.* 391, 923.
130. Grimm, H.-J., Gilfanov, M. and Sunyaev, R. 2003, *Mon. Not. R. Astron. Soc.* 339, 793.
131. Grindlay, J.E., Band, D., Seward, F., Leahy, D., Weisskopf, M.C. and Marshall, F.E. 1984, *Astrophys. J.* 277, 286.

132. Henson, G., Kemp, J. and Kraus, D. 1982, *IAU Circ.* N 3750.
133. Hirai, Y. and Fukue, J. 2001, *Publ. Astron. Soc. Japan* 53, 679.
134. Hjellming, R.M. and Johnston, K.J. 1981, *Astrophys. J.* 246, L141.
135. Hjellming, R.M. and Johnston, K.J. 1988, *Astrophys. J.* 328, 600.
136. Ho, L.C., Filippenko, A.V. and Sargent, W.L.W. 1997, *Astrophys. J.* 487, 658.
137. Holt, S.S., Schlegel, E.M., Hwang, U. and Petre, R. 2003, *Astrophys. J.* 588, 792.
138. Humphreys, R.M., and Davidson, K. 1994, *Publ. Astr. Soc. Pac.* 106, 1025.
139. Hut, P. and van den Heuvel, E.P.J. 1981, *Astron. Astrophys.* 94, 327.
140. Hynes, R.I., Clark, J.S., Barsukova, E.A., Callanan, P.J., Charles, P.A., Collier Cameron, A., Fabrika, S.N., Garcia, M.R., Haswell, C.A., Horne, K., Miroshnichenko, A., Negueruela, I., Reig, P., Welsh, W.F. and Witherick, D.K. 2002, Astron. Astrophys. 392, 991.
141. Icke, V. 1989, *Astron. Astrophys.* 216, 294.
142. Inoue, H., Shibazaki, N. and Hoshi, R. 2001, *Publ. Astron. Soc. Japan* 53, 127.
143. Irsmambetova, T.R. 1997, *Astron. Lett.* 23, 299.
144. Irsmambetova, T.R. 2001, *Astrophysics* 44, 243.
145. Jaroszynski, M., Abramowicz, M.A. and Paczynski, B. 1980, *Acta Astron.* 30, 1.
146. Johnston, K.J., Santini, N.J., Spencer, J.H., Klepczynski, W.J., Kaplan, G.H., Josties, F.J., Angerhofer, P.E., Florkowski, D.R. and Matsakis, D.N. 1981, *Astron. J.* 86, 1377.
147. Johnston, K.J., Geldzahler, B.J., Spencer, J.H., Waltman, E.B., Klepczynski, W.J., Josties, F.J., Angerhofer, P.E., Florkowski, D.R., McCarthy, D.D. and Matsakis, D.N. 1984, *Astron. J.* 89, 509
148. Jowett, F.H. and Spencer, R.E. 1995, in *"Proc. 27th YERAC"*, D.A. Green and W. Steffen (eds). Cambridge University Press, Cambridge, p.12
149. Kaaret, P., Corbel, S. Prestwich, A.H., Zezas, A. 2003, *Science* 299, 365.
150. Karachentsev, I.D., Sharina, M.E., Makarov, D.I., Dolphin, A.E., Grebel, E.K., Geisler, D., Guhathakurta, P., Hodge, P.W., Karachentseva, V.E., Sarajedini, A. and Seitzer, P. 2002, *Astron. Astrophys.* 389, 812.
151. Katz, J.I. 1980, *Astrophys. J.* 236, L127.
152. Katz, J.I. 1986, *Comments Astrophys.* 11, 201.
153. Katz, J.J. 1987, *Astrophys. J.* 317, 264.
154. Katz, J.I., Anderson, S.F., Grandi, S.A. and Margon, B. 1982, *Astrophys. J.* 260, 780.
155. Kawai, N., Matsuoka, M., Pan, H. and Stewart, G.C. 1989, *Publ.*

Astron. Soc. Japan 41, 491.
156. Kemp, J.C., Henson, G.D., Kraus, D.J., Carroll, L.C., Beardsley, I.S., Takagishi, K., Jugaku, J., Matsuoka, M., Leibowitz, E.M., Mazeh, T. and Mendelson, H. 1986, Astrophys. J. 305, 805.
157. Kilgard, R.E., Kaaret, P., Krauss, M.I., Prestwich, A.H., Raley, M.T and Zezas, A. 2002, Astrophys. J. 573, 138.
158. Kim, D,-W. and Fabbiano, G. 2003, Astrophys. J. 586, 826.
159. King, A.R., Davies, M.B., Ward, M.J., Fabbiano, G. and Elvis, M. 2001, Astrophys. J. 552, L109.
160. Kirshner, R.P. and Chevalier, R.A. 1980, Astrophys. J. 242, L77.
161. Kodaira, K., Nakada, Y. and Backman, D.E. 1985, Astrophys. J. 296, 232.
162. Koerding, E., Falcke, H., Markoff, S. and Fender, R. 2001, Astron. Gesells. Meet. Abstr. 18, 176.
163. Königl, A. 1983, Mon. Not. R. Astron. Soc. 205, 471.
164. Kopylov, I.M., Kumaigorodskaya, R.N. and Somova, T.A. 1985, Sov. Astron. 29, 186.
165. Kopylov, I.M., Kumaigorodskaya, R.N., Somov, N.N., Somova, T.A. and Fabrika, S.N. 1986, Sov. Astron. 30, 408.
166. Kopylov, I.M., Kumaigorodskaya, R.N., Somov, N.N., Somova, T.A. and Fabrika, S.N. 1987, Sov. Astron. 31, 410.
167. Kopylov, I.M., Bychkova, L.V., Fabrika, S.N., Kumaigorodskaya, R.N. and Somova, T.A. 1989, Sov. Astron. Lett. 15, 474.
168. Kotani, T., Kawai, N., Aoki, T., Doty, J., Matsuoka, M., Mitsuda, K., Nagase, F., Ricker, G. and White, N.E. 1994, Publ. Astron. Soc. Japan 46, L147.
169. Kotani, T., Kawai, N., Matsuoka, M. and Brinkmann, W. 1996, Publ. Astron. Soc. Japan 48, 619.
170. Kotani, T., Kawai, N., Matsuoka, M. and Brinkmann, W. 1997a, in "X-ray Imaging and Spectroscopy of Cosmic Hot Plasmas", F. Makino and K. Mitsuda (eds). Universal Academy Press, Tokyo, p.443.
171. Kotani, T., Kawai, N., Matsuoka, M. and Brinkmann, W. 1997b, in "Accretion Phenomena and Related Outflows". IAU Coll. N 163, D.T. Wickramasinghe, G.V. Bicknell and L. Ferrario (eds). Astron. Soc. of the Pacific, San Francisco, p.370.
172. Kotani, T. 1998, PhD. The Institute of Space and Astronautical Sciences. Japan.
173. Kotani, T., Kawai, N., Matsuoka, M. and Brinkmann, W. 1998, in "The Hot Universe". Proc. of IAU Symp. N 188. K. Koyama, S. Kitamoto, M. Itoh (eds). Kluwer Acad. Press, Dordrecht, p.358.
174. Kotani, T., Trushkin, S. and Denissyuk, E.K. 2002, In 'New Views on Microquasars", the Fourth Microquasars Workshop, Ph. Durouchoux, Y. Fuchs, J. Rodriguez (eds). Center for Space Physics, Kolkata (India), p. 257; astro-ph/0208250.

175. Kotoku, J., Mizuno, T., Kubota, A. and Makishima, K. 2000, *Publ. Astr. Soc. Japan* 52, 1081.
176. Kubota, A., Mizuno, T., Makishima, K., Fukazawa, Y., Kotoku, J., Ohnishi, T. and Tashiro, M. 2001, *Astrophys. J.* 547, L119.
177. Kubota, A., Done, C. and Makishima, K. 2002, *Mon. Not. R. Astr. Soc.* 337, L11.
178. La Parola, V., Peres, G., Fabbiano, G., Kim, D. W. and Bocchino, F. 2001, *Astrophys. J.* 556, 47.
179. Lebedev, S.V., Chittenden, J.P., Beg, F.N., Bland, S.N., Ciardi, A., Ampleford, D., Hughes, S., Haines, M.G., Frank, A., Blackman, E.G. and Gardiner, T. 2002, *Astrophys. J.* 564, 113.
180. Leibowitz, E.M. and Mendelson, H. 1982, *Publ. Astr. Soc. Pac.* 94, 977.
181. Leibowitz, E.M. 1984, *Mon. Not. R. Astron. Soc.* 210, 279.
182. Leibowitz, E.M., Mazeh, T., Mendelson, H., Kemp, J.C., Barbour, M.S., Takagishi, K., Jugaku, J. and Matsuoka, M. 1984, *Mon. Not. R. Astron. Soc.* 206, 751.
183. Liebert, J., Angel, J.R.P., Hege, E.K., Martin, P.G. and Blair, W.P. 1979, *Nature* 279, 384.
184. Lind, K.R. and Blandford, R.D. 1985, *Astrophys. J.* 295, 358.
185. Lipunov, V.M. and Shakura, N.I. 1982, *Sov. Astron.* 26, 386.
186. Lipunov, V.M., Ozernoy, L.M., Popov, S.B., Postnov, K.A. and Prokhorov, M.E. 1996, *Astrophys. J.* 466, 234.
187. Lipunova, G.V. 1999, *Astron. Lett.* 205, 508.
188. Liu, J.-F., Bregman, J.N., Irwin, J. and Seitzer, P. 2002a, *Astrophys. J.* 581, L93.
189. Liu, J.-F., Bregman, J.N. and Seitzer, P. 2002b, *Astrophys. J.* 580, L31.
190. Lubow, S.H. and Shu, F.H. 1975, *Astrophys. J.* 198, 383.
191. Lyndel-Bell, D. 1978, *Phys. Scripta* 17, 185.
192. Madau, P. 1988, *Astrophys. J.* 327, 116.
193. Madau, P. and Rees, M.J. 2001, *Astrophys. J.* 551, L27.
194. Makishima, K., Ohashi, T., Hayashida, K., Inoue, H., Koyama, K., Takano, S., Tanaka, Y., Yoshida, A., Turner, M.J.L., Thomas, H.D., Stewart, G.C., Williams, R.O., Awaki, H., Tawara, Y. 1989, *Pub. Astr. Soc. Japan* 41, 697.
195. Makishima, K., Kubota, A., Mizuno, T., Ohnishi, T., Tashiro, M., Aruga, Y., Asai, K., Dotani, T., Mitsuda, K., Ueda, Y., Uno, S., Yamaoka, K., Ebisawa, K., Kohmura, Y. and Okada, K. 2000, *Astrophys. J.* 535, 632.
196. Mammano, A. and Vittone, A. 1978, *IAU Circ.* N 3308, 3.
197. Margon, B. 1979, *IAU Circ.* N 3345, 1.
198. Margon, B., Grandi, S. and Ford, H. 1979a, *Bull. Am. Astron. Soc.* 11, 446.
199. Margon, B., Stone, R.P.S., Klemola, A., Ford, H.C., Katz, J.I.,

Kwitter, K.B. and Ulrich, R.K. 1979b, *Astrophys. J.* 230, L41.
200. Margon, B., Grandi, S.A., Stone, R.P.S. and Ford, H.C. 1979c, *Astrophys. J.* 233, L63.
201. Margon, B. and Anderson, S.F. 1989, *Astrophys. J.* 347, 448.
202. Margon, B. 1984, *Ann. Rev. Astron. Astrophys.* 22, 507.
203. Markoff, S., Falcke, H., and Fender, R. 2001, *Astron. Astrophys.* 372, L25.
204. Marshall, F.E., Mushotzky, R.F., Boldt, E.A., Holt, S.S. and Serlemitsos, P.J. 1978, *IAU Circ.* N 3314, 2.
205. Marshall, H.L., Canizares, C.R. and Schulz, N.S. 2002, *Astrophys. J.* 564, 941.
206. Matese, J.J. and Whitmire, D.P. 1982, *Astron. Astrophys.* 106, L9.
207. Matese, J.J. and Whitmire, D.P. 1983, *Astrophys. J.* 266, 776.
208. Matese, J.J. and Whitmire, D.P. 1984, *Astrophys. J.* 282, 522.
209. Mazeh, T., Leibowitz, E.M. and Lahav, O. 1981, *Astrophys. Lett.* 22, 55.
210. Mazeh, T., Aguilar, L.A., Treffers, R.R., Königl A. and Sparke, L.S. 1983, *Astrophys. J.* 265, 235.
211. Mazeh, T., Kemp, J.C., Leibowitz, E.M., Meningher, H. and Mendelson, H. 1987, *Astrophys. J.* 317, 824.
212. McAlary, C.W. and McLaren, R.A. 1980, *Astrophys. J.* 240, 853.
213. McCollough, M.L., Robinson, C.R., Zhang, S.N., Harmor, V.A., Hjellming, R.M., Waltman, E.B., Foster, R.S., Ghiggo, F.D., Briggs, M.S., Pendleton, G.N. and Johndton K.J. 1999, *Astrophys. J.* 517, 951.
214. McLean, I.S. and Tapia, S. 1980, *Nature* 287, 703.
215. Migliari, S., Fender, R. and Mendez, M. 2002, *Science* 297, 167.
216. Milgrom, M. 1979a, *Astron. Astrophys.* 76, L3.
217. Milgrom, M. 1979b, *Astron. Astrophys.* 78, L9.
218. Milgrom, M. 1981, *Vistas Astron.* 25, 141.
219. Miller, B.W. 1995, *Astrophys. J.* 446, L75.
220. Miller, M.C. and Hamilton, D.S. 2002, *Mon. Not. R. Astron. Soc.* 330, 232.
221. Miller, M.C. and Colbert, E.J.M. 2003, astro-ph/0308402.
222. Miller, J.M., Fabbiano, G., Miller, M.C. and Fabian, A.C. 2003a, *Astrophys. J.* 585, L37.
223. Miller, J.M., Zezas, A., Fabbiano, G. and Schweizer, F. 2003b, astro-ph/0302535.
224. Miller, J.M., Fabian, A.C. and Miller, M.C. 2003c, astro-ph/0310617.
225. Mirabel, I.F., Rodriguez, L.F., Cordier B., Paul, J. and Lebrun, F. 1992, *Nature* 358, 215.
226. Mirabel, I.F., Dhavan, V., Chaty, S., Rodriguez, L.F., Marti, J., Robinson, C.R., Swank, J. and Geballe, T. 1998, *Astron. Astrophys.* 330, L9

THE JETS IN SS433 147

227. Mirabel, I.F. and Rodriguez, L.F. 1999, *Ann. Rev. Astron. Astrophys.* 37, 409.
228. Mirabel, I.F. 2001, *Astrophys. and Space Sci. Suppl.* 276, 319.
229. Mizuno, T., Kubota, A. and Makishima, K. 2001, *Astrophys. J.* 554, 1282.
230. Molteni, D., Lanzafame, G. and Chakrabarti, S.K. 1994, *Astrophys. J.* 425, 161.
231. Mukai, K., Pence, W.D., Snowden, S.L. and Kuntz, K.D. 2003, *Astrophys. J.* 582, 184.
232. Murdin, P., Clark, D.H. and Martin, P.G. 1980, *Mon. Not. R. Astron. Soc.* 193, 135.
233. Namiki, M., Kawai, N., Kotani, T., Mamauchi, S. and Brinkmann, W. 2000, *Adv. Space Res.* 25, 709.
234. Namiki, M., Kawai, N., Kotani, T. and Makishima, K. 2003, *Publ. Astron. Soc. Japan* 55, 281.
235. Narayan, R., Nityananda, R. and Wiita, P.J. 1983, *Mon. Not. R. Astron. Soc.* 205, 1103.
236. Niell, A.E., Preston, R.A. and Lockhart, T.G. 1981, *Astrophys. J.* 250, 248.
237. Okada, K., Dotani, T., Makishima, K., Mitsuda, K. and Mihara, T. 1998, *Publ. Astr. Soc. Japan* 50, 25.
238. Okuda, T. and Fujita, M. 2000, *Publ. Astron. Soc. Japan* 52, L5.
239. Okuda, T. 2002, *Publ. Astron. Soc. Japan*, 54, 253.
240. Orosz, J.A. and Bailyn, C.D. 1997, *Astroph. J.* 477, 876.
241. Paczynsky, B. and Wiita, P. 1980, *Astron. Astrophys.* 88, 23.
242. Pakull, M.W. and Mirioni, L. 2001, *Astron. Gesells. Meet. Abstr.* 18, 12.
243. Pakull, M.W. and Mirioni, L. 2003, *Rev. Mex. de Astron. y Astrof. (Serie de Conferencias)* 15, 197.
244. Panferov, A.A. and Fabrika, S.N. 1993, *Astron. Lett.* 19, 41.
245. Panferov, A.A. and Fabrika, S.N. 1997, *Astron. Rep.* 41, 506.
246. Panferov, A.A., Fabrika, S.N. and Rakhimov, V.Yu. 1997, *Astron. Rep.* 41, 342.
247. Panferov, A.A. 1999, *Astron. Astrophys.* 351, 156.
248. Papaloizou, J.C.B. and Pringle, J.E. 1982, *Mon. Not. R. Astron. Soc.* 200, 49.
249. Papaloizou, J.C. and Pringle, J.E. 1983, *Mon. Not. R. Astron. Soc.* 202, 1181.
250. Paragi, Z., Vermeulen, R.C., Fejes, I., Schilizzi, R.T., Spencer, R.E. and Stirling, A.M. 1999, *Astron. Astrophys.* 348, 910.
251. Paragi, Z., Fejes, I., Vermeulen, R.C., Schilizzi, R.T., Spencer, R.E. and Stirling, A.M. 2000, In *"Galaxies and their Constituents at the Highest Angular Resolution"*. IAU Symp. N 205, R.T. Schilizzi (ed.), Manchester, United Kingdom, p.266.
252. Paragi, Z., Fejes, I., Vermeulen, R.C., Schilizzi, R.T., Spencer,

R.E. and Stirling, A.M. 2002, in *"6th VLBI Network" Symposium*, E. Ros, R.W. Porcas, A.P. Lobanov and J.A. Zensus (eds), p.263; astro-ph/0207061.
253. Pekarevich, M., Piran, T. and Shaham, J. 1984, *Astrophys. J.* 283, 295.
254. Peter, W. and Eichler, D. 1993, *Astrophys. J.* 417, 170.
255. Peter, W. and Eichler, D. 1996, *Astrophys. J.* 466, 840.
256. Petterson, J.A. 1981, *Adv. Space Res.* 1, 49.
257. Postnov, K.A. 2003, *Astron. Lett.* 29, 1.
258. Poutanen, J. and Zdziarski, A.A. 2002, in *"New Views on Microquasars"*, the Fourth Microquasars Workshop, Ph. Durouchoux, Y. Fuchs, and J. Rodriguez (eds). Center for Space Physics, Kolkata (India), p.268; astro-ph/0209186.
259. Rees, M.J., Phinney, E.S., Begelman, M.C. and Blandford, R.D. 1982, *Nature* 295, 17.
260. Revnivtsev, M., Sunyaev, R., Gilfanov, M. and Churazov, E. 2002a, *Astron. Astrophys.* 385, 904.
261. Revnivtsev, M., Gilfanov, M., and Churazov, E. and Sunyaev, R. 2002b, *Astron. Astrophys.* 391, 1013.
262. Roberts, W.J. 1974, *Astrophys. J.* 187, 575.
263. Roberts, T.R. and Warwick, R.S. 2000, *Mon. Not. R. Astron. Soc.* 315, 98.
264. Roberts, T.P. and Colbert, E.J.M. 2003, *Mon. Not. R. Astron. Soc.* 341, L49.
265. Roberts, T.P., Goad, M.R., Ward, M.J., Warwick, R.S., O'Brien, P.T., Lira, P. and Hands, A.D.P. 2001, *Mon. Not. R. Astron. Soc.* 325, L7.
266. Roberts, T.P., Goad, M.R., Ward, M.J. and Warwick, R.S. 2003 *Mon. Not. R. Astron. Soc.* 342, 709.
267. Romney, J.D., Schilizzi, R.T., Fejes, I. and Spencer, R.E. 1987, *Astrophys. J.* 321, 822.
268. Rowell, G.P. 2001, astro-ph/0104288.
269. Safi-Harb, S. and Oegelman, H. 1997, *Astrophys. J.* 483, 868.
270. Safi-Harb, S. and Petre, R. 1999, *Astrophys. J.* 512, 784.
271. Safi-Harb, S. and Kotani, T. 2002, in *"New Views on Microquasars"*, the Fourth Microquasars Workshop, Ph. Durouchoux, Y. Fuchs, and J. Rodriguez (eds). Center for Space Physics, Kolkata (India), 271; astro-ph/0210396.
272. Sarazin, C.L., Irwin, J.A. and Bregman, J.N. 2001, *Astrophys. J.* 556, 533.
273. Sawada,K., Matsuda, T. and Hachisu, I. 1986. *Mon. Not. R. Astron. Soc.* 221, 679.
274. Shapiro, P.R., Milgrom, M. and Rees, M.J. 1986, *Astrophys. J. Suppl. Ser.* 60, 393.
275. Seaquist, E.R., Gregory, P.C. and Crane, P.C. 1978, *IAU Circ.*

N 3256, 2.
276. Seaquist, E.R., Gilmore, W., Nelson, G.J., Payten, W.J. and Slee, O.B. 1980, *Astrophys. J.* 241, L77.
277. Seaquist, E.R. 1981, *Vistas Astron.* 25, 79.
278. Seaquist, E.R., Gilmore, W.S., Johnston, K.J. and Grindlay, J.E. 1982, *Astrophys. J.* 260, 220.
279. Seifina, E.V., Shakura, N.I., Postnov, K.A. and Prokhorov, M.E. 1991. *Lect. Notes Phys.* 385, 151.
280. Seward, F., Grindlay, J., Seaquist, E. and Gilmore, W. 1980, *Nature* 287, 806.
281. Shakura, N.I. 1972, *Sov. Astron.* 16, 756.
282. Shakura, N.I. and Sunyaev, R.A. 1973, *Astron. Astrophys.* 24, 337.
283. Shklovskii, I.S. 1960, *Sov. Astron.* 4, 243.
284. Shklovskii, I.S. 1981, *Sov. Astron.* 25, 315.
285. Sikora, M. 1981, *Mon. Not. R. Astron. Soc.* 196, 257.
286. Spencer, R. E. and Waggett, P. 1984, in *"VLBI and Compact Radio Sources"*. Proc.IAU Symp. N 110, R. Fanti (ed.), p.297.
287. Stephenson, C.B. and Sanduleak, N. 1977, *Astrophys. J. Suppl. Ser.* 33, 459.
288. Stewart, G.C., Watson, M.G., Matsuoka, M., Brinkmann, W., Jugaku, J., Takagishi, K., Omodaka, T., Kemp, J.C., Kenson, G.D., Kraus, D.J., Mazeh, T. and Leibowitz, E.M. 1987, *Mon. Not. R. Astron. Soc.* 228, 293.
289. Stone, J.M., Pringle, J.E. and Begelman, M.C. 1999, *Mon. Not. R. Astron. Soc.* 310, 1002.
290. Strohmayer, T.E. and Mushotzky, R.F. 2003, *Astrophys. J.* 586, L61.
291. Sugiho, M., Kotoku, J., Makishima, K., Kubota, A., Mizuno, T., Fukazawa, Y. and Tashiro, M. 2001, *Astrophys. J.* 561, L73.
292. Sunyaev, R. & Revnivtsev, M. 2000, *Astron. Astrophys.* 358, 617.
293. Trushkin, S.A., Bursov, N.N. and Smirnova, Yu.V. 2001, *Astron. Rep.* 45, 804.
294. Tsarevsky, G. 2002, private communication.
295. van den Heuvel, E.P.J., Ostriker, J.P. and Petterson, J.A. 1980, *Astron. Astrophys.* 81, L7.
296. van den Heuvel, E.P.J. 1981, *Vistas Astron.* 25, 95.
297. van der Laan, H. 1966, *Nature* 211, 1131.
298. van der Marel, R.P. 2003, astro-ph/0302101.
299. Velazquez, P.F. and Raga, A.C. 2000, *Astron. Astrophys.* 362, 780.
300. Vermeulen, R. C. 1996, in: *"Jets from stars and Galactic Nuclei"*, W. Kundt (ed.). Springer-Verlag, Berlin, Heidelberg, New York; also *Lect. Notes Phys.* 471, 122.
301. Vermeulen, R.C., Schilizzi, R.T., Icke, V., Fejes, I. and Spencer, R.E. 1987, *Nature* 328, 309.
302. Vermeulen, R.C., Murdin, P.G., van den Heuvel, E.P.J., Fabrika,

S.N., Wagner, B., Margon, B., Hutchings, J.B., Schilizzi, R.T., van Kerkwijk, M.H., van den Hoek, L.B., Ott, E., Angebault, L.P., Miley, G.K., D'Odorico, S. and Borisov, N. 1993a, *Astron. Astrophys.* 270, 204.
303. Vermeulen, R.C., Schilizzi, R.T., Spencer, R.E., Romney, J.D. and Fejes, I. 1993b, *Astron. Astrophys.* 270, 177.
304. Vermeulen, R.C., McAdam, W.B., Trushkin, S.A., Facondi, S.R., Fiedler, R.L., Hjellming, R.M., Johnston, K.J and Corbin, J. 1993c, *Astron. Astrophys.* 270, 189.
305. Voges, W., Aschenbach, B., Boller, Th., Braüninger, H., Briel, U., Burkert, W., Dennerl, K., Englhauser, J., Gruber, R., Haberl, F., Hartner, G., Hasinger, G., Kürster, M., Pfeffermann, E., Pietsch, W., Predehl, P., Rosso, C., Schmitt, J.H.M.M., Trümper, J., Zimmermann, H.U. 1999, Astron. Astrophys. 349, 389.
306. Voges, W., Aschenbach, B., Boller, Th., Braüninger, H., Briel, U., Burkert, W., Dennerl, K., Englhauser, J., Gruber, R., Haberl, F., Hartner, G., Hasinger, G., Pfeffermann, E., Pietsch, W., Predehl, P., Schmitt, J.H.M.M., Trümper, J., Zimmermann, H.U. 2000, http://wave.xray.mpe.mpg.de/rosat/catalogues/rass-fsc.
307. Wagner, R.M. 1986, *Astrophys. J.* 308, 152.
308. Wang, Q.D. 2002, *Mon. Not. R. Astron. Soc.* 332, 764.
309. Watson, M.G., Willingale, R., Grindlay, J.E. and Seward, F.D. 1983, *Astrophys. J.* 273, 688.
310. Watson, M.G., Stewart, G.C., King, A.R. and Brinkmann, W. 1986, *Mon. Not. R. Astron. Soc.* 222, 261.
311. Whitmire, D.P. and Matese, J.J. 1980, *Mon. Not. R. Astron. Soc.* 193, 707.
312. Wu, H., Xue, S.J., Xia, X.Y., Deng, Z.G. and Mao, S. 2002, *Astrophys. J.* 576, 738.
313. Yamauchi, S., Kawai, N. and Aoki, T. 1994. *Pub. Astr. Soc. Japan* 46, L109.
314. Yuan, W., Kawai, N., Brinkmann, W. and Matsuoka, M. 1995, *Astron. Astrophys.* 297, 451.
315. Zampieri, L., Mucciarelli, P., Falomo, R., Kaaret, P., Di Stefano, R., Turolla, R., Chieregato, M. and Treves, A. 2003, astro-ph/0309687.
316. Zealey, W.J., Dopita, M.A. and Malin, D.F. 1980, *Mon. Not. R. Astron. Soc.* 192, 731.
317. Zezas, A. and Fabbiano, G. 2002, *Astrophys. J.* 577, 726.
318. Zezas, A., Fabbiano, G., Rots, A.H. and Murray, S.S. 2002, *Astrophys. J.* 577, 710.
319. Zwitter, T., Calvani, M., Bodo, G. and Massaglia, S. 1989, *Fundam. Cosmic Phys.* 13, 309.
320. Zwitter, T., Calvani, M. and D'Odorico, S. 1991, *Astron. Astrophys.* 251, 92.

INDEX

Balmer decrements 61, 79, 98
B[e]-supergiant 128
bullets 18, 19, 22, 34

clouds 7, 18, 21–26, 34–35, 55–65, 78–79
cloudlets 23
cocoon 53, 65, 75, 82, 87, 109, 122
collimated radiation 48, 65, 66, 69–74, 79, 128, 132–135
critical Roche lobe 3, 27, 31, 35, 42, 82, 105
crossover 12, 13

donor star 3, 4, 6, 27, 42, 84, 87–90, 94, 98, 100, 105–107, 113, 118, 120

eclipses 4, 31, 43, 80–87, 105, 112
— primary, Min I 80–86, 94, 103, 107, 109, 113
— secondary, Min II 80–86
— X-ray 43, 44, 46, 53, 84, 109
ephemerides 16, 29, 84, 85, 91, 111
equatorial wind 35–37, 48, 49, 116–120

filling factor 55, 62, 63, 67, 136
flares 6, 18, 29–32, 81–82, 86
— optical 18, 30, 31, 78–80, 86
— radio 27, 29–32
funnel 23, 30, 31, 56, 65, 66, 69–78, 87, 131, 132, 134

intermediate mass black hole (IMBH) 131, 136

jitter 17, 31, 52

kinematic model 7, 12–16, 32, 36, 38, 40, 50, 114

de Laval nozzle 72

line locking 74–75
Lorentz factor 13
luminosity 5, 90
— bolometric 5, 43, 90, 95, 97, 110
— emission lines 62, 65, 100
— critical 70, 71, 94, 115, 127, 128
— kinetic 7, 39, 44, 47, 52, 62, 68, 80, 90
— radio 7
— X-ray 5, 40, 41, 43, 45, 47, 52, 65, 90
luminosity function 126, 127, 129, 130
luminous blue variable (LBV) 100

microquasar 8, 9, 24, 122–126, 129, 132, 134
masses 101, 104, 106, 111
— mass function 101–106
— mass ratio 82, 104–106

nodding motion 8, 27–29, 31, 84, 87, 88, 92

optical filaments 7, 38, 40, 65
orbital
— alignment 8, 27, 42
— eccentricity 31, 32, 88
— circularization 42
— period variation 16, 105

P Cyg line profile 6, 10, 101, 111, 113
Paschen lines 110, 111, 118
phases 12, 43, 84, 85, 88, 94
— precessional 12–14, 43, 85
— orbital 43, 85, 91, 102, 103
— nutational 19, 85, 88
photosphere 6, 26, 31, 46, 50, 53, 54, 64, 77, 82, 94, 96, 97, 100, 115, 120, 131
precession 8, 15
— driven 8, 15, 82, 88, 118
— slaved 8, 18, 27, 42, 84, 88, 118

polarization 98, 99
— optical 67, 98, 99
— radio 26
— UV 87, 98, 99

radial velocity curve 101–103
relativistic aberration 43, 58, 60
relativistic beaming 28, 48, 131
— slaved disk 42, 82
— slim disk 129

spherization radius 71, 72, 114
spiral shocks 23, 93, 112
state 6, 16, 27, 86, 87, 96
— active 6, 16, 27, 28, 31, 56, 86, 87, 94, 97, 123, 125
— quiescent 6, 16, 28, 29, 86, 87, 124
— low/hard 124, 125
Supernova 37, 38, 42
— asymmetrical SN explosion 42, 43
— SN remnant 7, 37, 41, 130

thermal instabilities 18, 23, 54, 69, 76–79
transverse Doppler effect 13, 38

ultraluminous X-ray sources (ULX) 122, 129–136

X-ray binaries 4, 122
— CI Cam 128
— Cyg X-1 122, 132
— Cyg X-3 125, 126
— GRO J1655-40 123
— GRS 1915+105 123, 124
— Her X-1 16
— Sco X-1 122
— V4641 Sgr 128

WR (WN, WNL) star 100

zone 35, 61
— radio brightening 32–35, 45, 46, 67
— sweep up 62–65
— expansion 35, 50, 62–65

www.ingramcontent.com/pod-product-compliance
Ingram Content Group UK Ltd.
Pitfield, Milton Keynes, MK11 3LW, UK
UKHW041418180426
11947UKWH00007B/199